The Hong Kong Economic Policy Studies Series

# COMPETITION IN ENERGY

# COMPETITION IN ENERGY

Pun-Lee Lam

Published for
The Hong Kong Centre for Economic Research
The Hong Kong Economic Policy Studies Forum
by

City University of Hong Kong Press

First published 1997
Printed in Hong Kong

ISBN 962-937-005-0

Published by
City University of Hong Kong Press
City University of Hong Kong
Tat Chee Avenue, Kowloon, Hong Kong

Internet: http://www.cityu.edu.hk/upress/
E-mail: upress@cityu.edu.hk

The free-style calligraphy on the cover, *neng*, means "*energy*" in Chinese.

# Contents

# Detailed Chapter Contents

# Foreword

The key to the economic success of Hong Kong has been a business and policy environment which is simple, predictable and transparent. Experience shows that prosperity results from policies that protect private property rights, maintain open and competitive markets, and limit the role of the government.

The rapid structural change of Hong Kong's economy in recent years has generated considerable debate over the proper role of economic policy in the future. The impending restoration of sovereignty over Hong Kong from Britain to China has further complicated the debate. Anxiety persists as to whether the pre-1997 business and policy environment of Hong Kong will continue.

During this period of economic and political transition in Hong Kong, various interested parties will be re-assessing Hong Kong's existing economic policies. Inevitably, some will advocate an agenda aimed at altering the present policy making framework to reshape the future course of public policy.

For this reason, it is of paramount importance for those familiar with economic affairs to reiterate the reasons behind the success of the economic system in the past, to identify what the challenges are for the future, to analyze and understand the economy sector by sector, and to develop appropriate policy solutions to achieve continued prosperity.

In a conversation with my colleague Y. F. Luk, we came upon the idea of inviting economists from universities in Hong Kong to take up the challenge of examining systematically the economic policy issues of Hong Kong. An expanding group of economists (The Hong Kong Economic Policy Studies Forum) met several times to give form and shape to our initial ideas. The Hong Kong Economic Policy Studies Project was then launched in 1996 with some 30 economists from the universities in Hong Kong and a few from overseas. This is the first time in Hong Kong history that a concerted public effort has been undertaken by academic economists in the territory. It represents a joint expression of our collective concerns, our hopes for a better Hong Kong, and our faith in the economic future.

The Hong Kong Centre for Economic Research is privileged to be co-ordinating this Project. The unfailing support of many distinguished citizens in our endeavour and their words of encouragement are especially gratifying. We also thank the directors and editors of the City University of Hong Kong Press and The Commercial Press (H.K.) Ltd. for their enthusiasm and dedication which extends far beyond the call of duty.

Yue-Chim Richard Wong
Director
The Hong Kong Centre
for Economic Research

# Foreword by Series Editor

Despite the small physical size of Hong Kong, its energy sector is quite colourful. Suppliers of the two main sources of energy, electricity and gas, have their corresponding market power and market shares, but are subject to different schemes of government regulation. Interestingly, some energy companies also involved in business not directly related to the energy industry. This peculiar situation presents an interesting case not only for policy making in Hong Kong but also for general economic analysis of competition and regulation.

It is obvious that the general public are concerned with the price of energy and the quality of services of the suppliers; for, the supply conditions of energy have direct bearing on each and every household and business. In recent years when Hong Kong experiences relatively high rates of inflation, the public have become less and less tolerant of hikes in energy prices. As various population subgroups are more and more represented in the legislative structure, their voices have to be reckoned by the government in designing regulatory policies on energy suppliers. Yet, would the public, being consumers of energy, overlook other relevant issues when they defend their consumer interests?

What are the overall framework and general policies of Hong Kong government's regulation of energy suppliers? Why are there different control schemes for different suppliers? How have these schemes evolved over time, and how do Hong Kong's schemes compare with regulation policies in other economies? Is the current arrangement in Hong Kong more beneficial to consumers, or to the energy investors? All these issues have to be clarified before one can have a good understanding of the current energy industry and come up with objective and appropriate policy proposals.

The most prominent feature of the energy industry is the huge capital cost required and the economies of scale, two conditions that would most likely lead to the situation of "natural monopolies". As such, the energy industry cannot completely avoid government intervention. We should keep in mind though, that although government regulation could promote competition and benefit the consumers, intervention could also impede competition and cause harm to the interest of consumers. The impacts of each regulatory policy must be carefully analyzed.

At the same time when advances in science and technology drastically can shape the supply and market structure of energy, economic growth and social development can shift the consumers' demand of energy from one kind of energy to another. Thus, the regulatory framework may require constant review and appropriate amendments to suit new environments.

This book presents a succinct discussion of the evolution and current situation of the electricity and gas markets in Hong Kong. It also analyzes regulatory policies currently in place and their relative advantages and disadvantages vis-à-vis those of other economies. Towards the end of the study, the author comes up with several concrete policy proposals to enhance competition in Hong Kong's energy industry.

The author, Dr. Pun-lee Lam, is an economist with strong research interests in the regulation of public utilities. He has studied and written widely about public utilities in Hong Kong. This book provides ample background materials and objective analysis for the discussion of energy policies, and is a worthy reference work for the concerned public.

Y. F. Luk
School of Economics and Finance
The University of Hong Kong

# Preface

I would like to express my gratitude to two anonymous referees for their generous and enlightening comments and advice. Their suggestions have greatly improved the structure of this study and expanded my knowledge of the field.

Throughout my study I received kind support from the electricity and gas companies in Hong Kong. They furnished me with the requested information and made every effort to check their company records and documents. I would like to thank K. Y. Mak and Karen Chan, both of them helped me obtain useful data and documents on the energy industry in Hong Kong. Finally, I want to thank Christine K. W. Loh for providing information about the operations of two electricity companies, and the Chinese University Press for permitting me to use information in its publication.

Pun-Lee Lam
Department of Business Studies
The Hong Kong Polytechnic University

# List of Illustrations

## Figures



## Tables

# Acronyms and Abbreviations

| Acronyms | Names in Full | Appear First on Page |
|---|---|---|
| BG | British Gas | 17 |
| BGE | British Gas Energy | 76 |
| BT | British Telecommunications | 74 |
| CAPCO | Castle Peak Power Company Limited | 10 |
| CAPM | Capital asset pricing model | 44 |
| CEGB | Central Electricity Generating Board | 71 |
| CLP | China Light and Power Company Limited | 2 |
| DGES | Director General of Electricity Supply | 72 |
| EPAct | Energy Policy Act | 64 |
| EPD | Environmental Protection Department | 29 |
| ESI | Electricity Supply Industry | 71 |
| EWGs | Exempt wholesale generators | 64 |
| FERC | Federal Energy Regulatory Commission | 63 |
| FPA | Federal Power Act | 62 |
| GGPC | Guangdong General Power Company | 11 |
| GNIC | Guangdong Nuclear Investment Company | 12 |
| GNPJVC | Guangdong Nuclear Power Joint Venture Company | 12 |
| HEC | Hongkong Electric Company Limited | 2 |
| HKCG | Hongkong and China Gas Company Limited | 2 |
| HKNIC | Hong Kong Nuclear Investment Company | 12 |
| IEA | International Energy Agency | 10 |
| KESCO | Kowloon Electricity Supply Company Limited | 10 |
| LNG | Liquefied natural gas | 18 |
| LPG | Liquefied petroleum gas | 2 |
| MMC | Monopolies and Mergers Commission | 72 |
| NGC | National Grid Company | 72 |

| | | |
|---|---|---|
| NGPA | National Gas Policy Act | 65 |
| OECD | Organisation for Economic Cooperation and Development | 41 |
| OFFER | Office of Electricity Regulation | 72 |
| OFGAS | Office of Gas Supply | 75 |
| OFT | Office of Fair Trading | 75 |
| OFTA | Office of the Telecommunications Authority | 97 |
| OPEC | Organisation of Petroleum Exporting Countries | 63 |
| PEPCO | Peninsula Electric Power Company Limited | 9 |
| PSDC | Hong Kong Pumped Storage Development Company Limited | 12 |
| PUHCA | Public Utility Holding Company Act | 62 |
| PURPA | Public Utility Regulatory Policies Act | 63 |
| RECs | Regional Electricity Companies | 72 |
| SNG | Substitute natural gas | 18 |

# Competition in Energy

CHAPTER 1

# Introduction

## Government Regulation of Industries

For decades economists and politicians around the world have often attributed Hong Kong's success to its dedication to a free enterprise system. It has been suggested that government intervention is minimal and that even if intervention is called for, it is for the purpose of correcting market failures and facilitating the smooth functioning of the free market. The notion of "positive non-intervention" has been used to justify the government's passive industrial policy.

Careful investigation, however, would reveal that industries in Hong Kong are not really free from the influence of the government's visible hand. Apart from direct involvement in housing, education and medical services, the government controls some major industries too. Transportation and public utilities are two typical examples. All railway systems in Hong Kong are owned and run by the government. Franchises have been granted to bus companies which enjoy monopoly rights in the provision of bus services. Taxi and minibus services are both subject to entry control and, with the exception of minibuses, all modes of transportation are under fare regulation. Except for gas supply, utility companies are under some form of direct government control. Water is supplied solely by the government, while electricity is supplied by two regional monopolists. Local and international telephone services have also been under franchised monopoly, subject to tight government control. Some of these utility companies have been subject to a rate-of-return regulation which is called the Scheme of

Control. Government intervention (or non-intervention) in these industries is thus a clear evidence of Hong Kong's interference with free market competition.

## Energy Industry in Hong Kong

This policy study examines the Hong Kong energy industry. In Hong Kong, electricity is provided by two regional monopolists, the China Light and Power Company Limited (CLP) and the Hongkong Electric Company Limited (HEC). These two electric utilities are governed by the Scheme of Control, which is a formal, long-term (fifteen-year) contract made between a private firm and the government. Under the Scheme of Control, a regulated utility is subject to both price control and (nominal) rate-of-return control.

Two main types of gas are available in Hong Kong, towngas and liquefied petroleum gas (LPG). Towngas is supplied solely by the Hong Kong and China Gas Company Limited (HKCG). The company manufactures towngas by using naphtha as a feedstock. On the other hand, LPG is supplied by six major oil companies based in Hong Kong. Four of these companies are: Shell Hong Kong Limited, Mobil Oil Hong Kong, Esso Hong Kong Limited, and Caltex Oil Hong Kong Limited. LPG supplies are delivered in two forms, cylinders and pipelines. LPG is imported into Hong Kong by sea. In 1995 about 62% of the total sales of LPG were distributed to customers in portable cylinders via dealer networks. There are about 522 LPG distributors operating within the city. The remaining 38% of LPG sales were distributed through piped gas systems from bulk LPG storage and vaporizer installations that are located in or adjacent to property developments that were supplied (*Hong Kong Annual Report* 1996).

Natural gas became available in Hong Kong at the end of 1995 and is used exclusively by CLP for power generation. In 1991 CLP won a bid over HKCG to purchase natural gas from a newly discovered gas field (Yacheng 13-1) off Hainan Island in China. The gas is transported directly from the gas field via a 778-kilometre, high-pressure submarine pipeline.

Thus there are distinct market structures under which energy companies operate. While the two electricity companies are regional monopolists under the Scheme of Control, oil companies are oligopolists. Although HKCG is a monopolist in the supply of towngas, without any government regulations on prices and returns, the firm still faces competition from other fuel suppliers.

## Industry Performance of Power Utilities

It has been alleged that these power utilities in Hong Kong, be they regulated or not, are efficiently run, that they have served the community well by providing Hong Kong with adequate power supply at a competitive price, and that infrastructure developed by these power companies has increased Hong Kong's competitiveness and helped Hong Kong develop over the past few decades. However, if we take a closer look at the prices charged by the utilities, and at their equity returns, there is evidence that government policies on these power utilities tend to protect the producers rather than the consumers.

Although the two electricity companies are governed by the Scheme of Control, they are able to earn returns well above their capital costs by debt financing. The unregulated towngas company has also earned attractive returns in recent years because there are barriers that limit market competition. The towngas company's market share in the gas industry has been expanding rapidly, causing public concerns over possible abuses of market power. Evidence shows that the price of electricity and towngas has not decreased in real terms over the last two decades. Although energy prices in Hong Kong are still reasonable, it could be lower because the city is highly developed and densely populated.

Despite technological improvement and capital investment over time, labour productivity of the two electricity companies has been growing very slowly. Although the labour productivity of the towngas company has shown remarkable improvement over time, this productivity gain is only reflected in higher returns for the company and has not been shared with customers in terms of a lower price. There has also been increasing public concern over higher energy prices resulting from the over-expansion of power utilities and unsatisfactory load management.

Since the two electricity companies are regulated while the towngas company is not, there have been complaints about unfair competition within the energy industry. On the one hand, the towngas company has argued that an electricity company can expand its market share in the fuel market by subsidizing certain customer groups in their purchase of electric appliances. The extra expenditure can then be automatically passed on to other customers. As the permitted returns of the company are based on capacity expansion, this cross-subsidizing strategy would increase the demand for electricity and subsequently the permitted returns of the company. On the other hand, the two electricity companies have claimed that their competition is limited by a lack of flexibility in pricing. The towngas company can use its pricing flexibility to capture a larger share of the market. Hence, an asymmetry in government policies on electricity and towngas companies may have hindered fair competition in the energy market. In other words, these alternative energy suppliers are not competing on a level playing field.

Even within the electricity industry, existing government policies are deficient in protecting the general interest of the consumers. Historically, HEC's electricity prices have been much higher than those of CLP. Customers on Hong Kong Island are unable to enjoy the lower prices charged by CLP. Such a large price differential would not persist if the two companies were in direct competition. Besides, these two electricity companies are regional monopolists, and their expansions are based on regional demands rather than on the total demand for electricity for the whole city. The Scheme of Control allows each company to build its own generation and transmission facilities, even though cheaper and better energy supplies can be easily obtained from other areas within or outside Hong Kong. This problem is clearly highlighted by HEC's recent plan to expand. In spite of a foreseeable power surplus by CLP in the first few years of the next century, the government may still approve HEC's expansion plan to meet the energy demand from Hong Kong Island during the same period. This lack of competition and co-operation between the two electricity companies thus results in a waste of resources for the whole society.

## The Scope of Study

Over the past decade, the energy industry in many countries has undergone substantial changes in organizational structure. Apart from the privatization of electric and gas utilities, competition has replaced monopoly in certain production stages of the industry. Against this current of competition, the structure of the energy industry in Hong Kong remains the same as it has been for decades. Drawing from the lessons gained by other countries, we suggest a tentative approach towards improving efficiency and competition within the energy industry in Hong Kong. The scope of this policy study is limited to government policies relating to the power utilities in Hong Kong and it does not cover energy policies on the transport sector. At present, two thirds of the primary energy requirements in Hong Kong are for producing. Hence, the scope of our study still covers a major portion of the entire energy sector.

The structure of this book is organized as follows. In Chapter 2, we introduce the background of the energy industry and the changing pattern of energy consumption in Hong Kong. In Chapter 3, we analyze government policies and the performance of power utilities in Hong Kong. Based on the experiences of other countries, which are discussed in Chapter 4, some proposals for enhancing competition in the energy industry in Hong Kong are outlined in Chapter 5. It will be shown that the market competition rather than regulation of monopoly can better protect consumer interests. This does not mean that competition and regulation are perfect substitutes. In some situations, government regulation is still needed to guard against anti-competitive behaviour by the dominant firms in a newly liberalized market. Hence, both competition and regulation (in certain production stages) are required to promote efficiency in the energy industry. Government policy should aim to remove barriers to competition. Chapter 6 is a summary of our recommendations.

# CHAPTER 2

# The Energy Industry in Hong Kong

## The Electricity Industry

Hong Kong's two electricity companies have had a long history in the city. The Hongkong Electric Company Limited (HEC), the first electricity company in Hong Kong, was incorporated in 1889 and began supplying electricity to Hong Kong Island in December 1890. The China Light and Power Company Limited (CLP) was incorporated in Hong Kong in 1901 for the purpose of supplying electricity to Canton (in China) and Kowloon. It was wound-up and incorporated again in 1918.

Until the early 1980s there was another smaller electricity company in Hong Kong called the Cheung Chau Electric Company Limited. This company was founded in 1913, originally as a community project, and it supplied electricity to the population and industries on Cheung Chau Island. In January 1984 the government authorized CLP to provide electricity to Cheung Chau. An agreement was reached between CLP and the Cheung Chau Electric Company under which CLP acquired the assets of the latter. Islanders have since benefited from lower electricity tariffs. Historically, there were minor enterprises such as village co-operatives that produced electricity to remote localities. After World War II, CLP expanded its services to cover these areas, and most of the co-operatives had been closed. Thus, electricity is currently supplied by two commercial regional monopolists: CLP and HEC.

## The Nationalization Fiasco of the 1960s

In 1952 CLP and HEC announced that a fuel surcharge of 17% would be imposed on all bills. The new surcharge was strongly opposed by industrialists, and the dissatisfaction among large consumers was highlighted in a petition sent to the governor by the Kowloon Chamber of Commerce. A committee of industrialists, mostly from Kowloon, formed with the intent to oppose the price increase. Another petition, led by the Chinese General Chamber of Commerce, called for a commission of inquiry, demanded state control of the two companies, and accused HEC of "destroying industry".

Amid public outcry Governor Sir Robert Black appointed on 16 July 1959 a three-man commission called the Electricity Supply Companies Commission. Two of the commissioners (one being a former chairman of the regional boards of Britain's nationalized electricity supply) were from England, and the third was from Hong Kong. The commission held a series of public hearings and submitted a report in December 1959. Instead of proposing any form of government regulation, the commission recommended an outright take-over of the two companies by the government in order to "permanently remove the competing interest between the shareholders and the public".

Negotiations between the government and the two companies then started in an attempt to devise some form of public control that was different from the compulsory public acquisition recommended by the commission. In November 1964, CLP, in co-operation with Esso Eastern Inc. (the trade name for the overseas operations of the parent company Exxon), proposed a Scheme of Control for fifteen years (lasting until September 1978) and eventually reached an agreement with the government.

The objectives of the scheme were to limit the disposable profits of the companies to a reasonable return on their equity capital while providing adequate incentives towards efficiency and expansion. In particular, the scheme was to ensure that the benefits from any capital for expansion obtained from additional profits would go primarily to the consumers. CLP promised to reduce tariffs and to set limits on dividends in subsequent years. These objectives were to be achieved by setting up a development fund. Any excess profits would be put into this fund and used to acquire fixed assets. The interest accrued would be used to reduce tariffs. CLP and the government agreed that the fund would constitute a liability, not an asset, of the company. Other important features of the joint-venture between CLP and Esso included:

1. The formation of a new generating company, the Peninsula Electric Power Company Limited (PEPCO), which would be jointly owned by CLP (40%) and Esso (60%). The electricity generated was to be sold exclusively to CLP and distributed through CLP's system.
2. Both CLP and PEPCO would, as soon as possible, purchase all their existing petroleum requirements at competitive prices from Esso under a long-term fuel supply contract.

Although HEC was not formally subject to the Scheme of Control until 1979 (when CLP renewed its contract with the government), the company maintained close co-operation with the government on all matters concerning tariff policy.

## Oil Price Shocks in the 1970s

In the 1970s the cost of fuel oil constituted a large part of the running costs of the two electricity companies, as their generating units mainly consumed fuel oil. In 1973 and 1974 the price of fuel oil rose sharply to about four times its former level, bringing about a substantial increase in the fuel cost adjustment portion of electricity tariffs. In addition, marked increases in other costs caused CLP's basic tariff to rise by one cent per unit in April 1974. This was the first increase in many years. Since fuel oil was in short supply, the government had to impose restrictions on the use of electricity during the first few months of 1974. These restrictions were lifted towards the end of May, once the supply situation improved.

In April 1979, owing to another fuel oil shortage, the government decided to implement oil conservation measures. The Oil (Conservation and Control) Ordinance 1979 conferred with the Governor in Council and the Director of Oil Suppliers to give directions to suppliers and dealers as to the storage, supply, use and disposal of oil, and to give similar directions to electricity and gas companies. A package of conservation measures was introduced in May 1979, but most parts had been lifted by October 1979, when the oil supply situation improved. To decrease dependence on oil and to help keep consumer costs down, the two electricity companies decided to construct new generators that would be capable of being fired by coal or oil. The decision was consistent with the International Energy Agency (IEA)'s 1977 policy that construction of exclusively oil-fired stations should be discouraged.

## Rapid Expansion in the 1980s

During the 1970s HEC diversified its activities into non-utility areas. In July 1976 HEC and other subsidiaries were grouped under an umbrella company, Hongkong Electric Holdings Limited. In April 1982 the Hongkong Land Company Limited acquired 34% of HEC's equity. In January 1985 the debt-ridden Hongkong Land decided to sell its holding to Hutchison Whampoa in order to reduce its debt-equity ratio. Then in March 1987 the Board of Directors of HEC proposed a reorganization in order to separate utility and non-utility activities. This reorganization was approved and became effective in June 1987.

There were no major structural changes in CLP, despite the fact that it had been expanding at a faster rate than had HEC in the preceding decades (see Figure 2.1), and continued to do so during the 1980s. In addition to the formation of PEPCO in 1964, two associated generating companies, Kowloon Electricity Supply Company Limited (KESCO) and Castle Peak Power Company Limited (CAPCO), were formed in 1978 and 1981, respectively. According to the previous arrangement, a 40% share of these two companies was held by CLP, and a 60% share was held by Exxon

**Figure 2.1**
**Electricity Consumption in Hong Kong, 1952–1995**

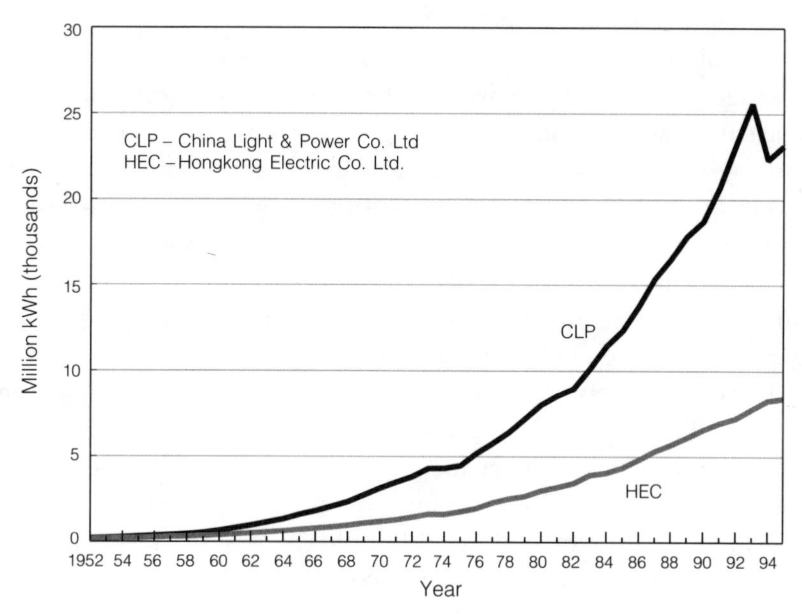


Energy Limited (a wholly owned subsidiary of Esso Eastern Incorporation). In April 1992 CAPCO acquired all the shares and undertakings of PEPCO and KESCO to become a single, strong, power-generating company with improved prospect's for future borrowing.

Lawrence Kadoorie, Chairman of CLP from 1956 to 1993, envisaged expanding the company's supply area to cover mainland China. This vision began to materialize in the 1980s.

The question of supplying electricity to Guangdong Province in the People's Republic of China was first raised when Kadoorie visited Beijing in May 1978. A series of meetings between CLP and the Guangdong Power Company (renamed Guangdong General Power Company (GGPC) in 1984) ensued, resulting in an agreement to interconnect the systems of the two companies and in CLP agreeing to supply electricity to Guangdong at a bulk tariff rate. In recent years, Guangdong's reliance on CLP's electricity supply has decreased as generating facilities in Southern China are commissioned.

In discussions on the interconnection between CLP and the Guangdong Power Company, the question of constructing a nuclear power station in Guangdong to supply electricity to Guangdong and Hong Kong was raised. After conducting several feasibility studies, the Chinese government announced its decision to proceed with the development of the nuclear power station at Daya Bay in late 1982. Later, the Hong Kong government also announced its willingness for Hong Kong to purchase electricity generated by the nuclear power station.

The Daya Bay station is now operated by the Guangdong Nuclear Power Joint Venture Company (GNPJVC), in which the Guangdong Nuclear Investment Company (GNIC) holds a 75% interest, and the Hong Kong Nuclear Investment Company (HKNIC), wholly owned by CLP, holds a 25% interest. The two companies signed the joint venture contract on 18 January 1985. It was agreed that about 70% of the power from this station would be supplied to Hong Kong. In spite of vehement opposition from the Hong Kong public, the government decided to proceed with the project. After some delays, the Daya Bay nuclear power station started its operation in mid 1993.

In December 1990 CLP, through its associate, Hong Kong Pumped Storage Development Company Limited (PSDC), reached an agreement for the right to use 50% of the pumped storage power station (capacity 1,200 MW), constructed at Conghua, north of Guangzhou (Canton). PSDC's shares are divided between Exxon Energy Limited (51%) and CLP (49%). Since the project is located outside Hong Kong, the investment is not governed by the Scheme of Control. Furthermore, CLP has maintained close links with other power companies in China through various feasibility studies and training courses.

At present, CLP's system is interconnected with that of HEC. Initially, the interconnection, commissioned in 1981, had a capacity of 480 MVA, but this was later increased to the current capacity of 720 MVA. The interconnection enhances the security of supply to both companies and produces cost savings by the economical interchange of electricity and the reduction of operating reserve requirements. In recent years, CLP and HEC have expanded their businesses into the property market by developing their former plant sites. A chart showing the structure of the electricity system in Hong Kong is provided in Figure 2.2.

## Concluding Remarks

In conclusion, we can divide the historical development of the Hong Kong electricity industry into three distinct periods. The first covers the period before 1964, when both companies were under no government control on price or profit. The second period is from 1964 to 1978 during which CLP was under the Scheme of Control. The third period is from 1979 onwards when both companies were under the Scheme of Control.

Before the early 1950s Hong Kong relied heavily on entrepôt trade for its income. The United Nations' embargo on Communist China, which was put in place when the Korean War broke out, forced Hong Kong to develop its own industries. Consequently, the demand for electricity, particularly the industrial demand, increased significantly. Industrialists played an important role in Hong Kong's industrialization process, and they were on the front line lobbying for government control of the two electricity companies.

During the second period Hong Kong's industries, together with the demand for electricity, continued to grow at a rapid pace. The growth momentum was hampered somewhat by the 1973 oil crisis, which caused the prices of fuel and electricity to skyrocket. The government subsequently introduced quantity controls on the use of electricity. Despite the shortages, Hong Kong's economy recovered rapidly in 1976.

In 1978 the Scheme of Control for CLP was renewed and China announced its open door policy. As a result more and more factories

**Figure 2.2**
**The Hong Kong Electricity System (1995)**

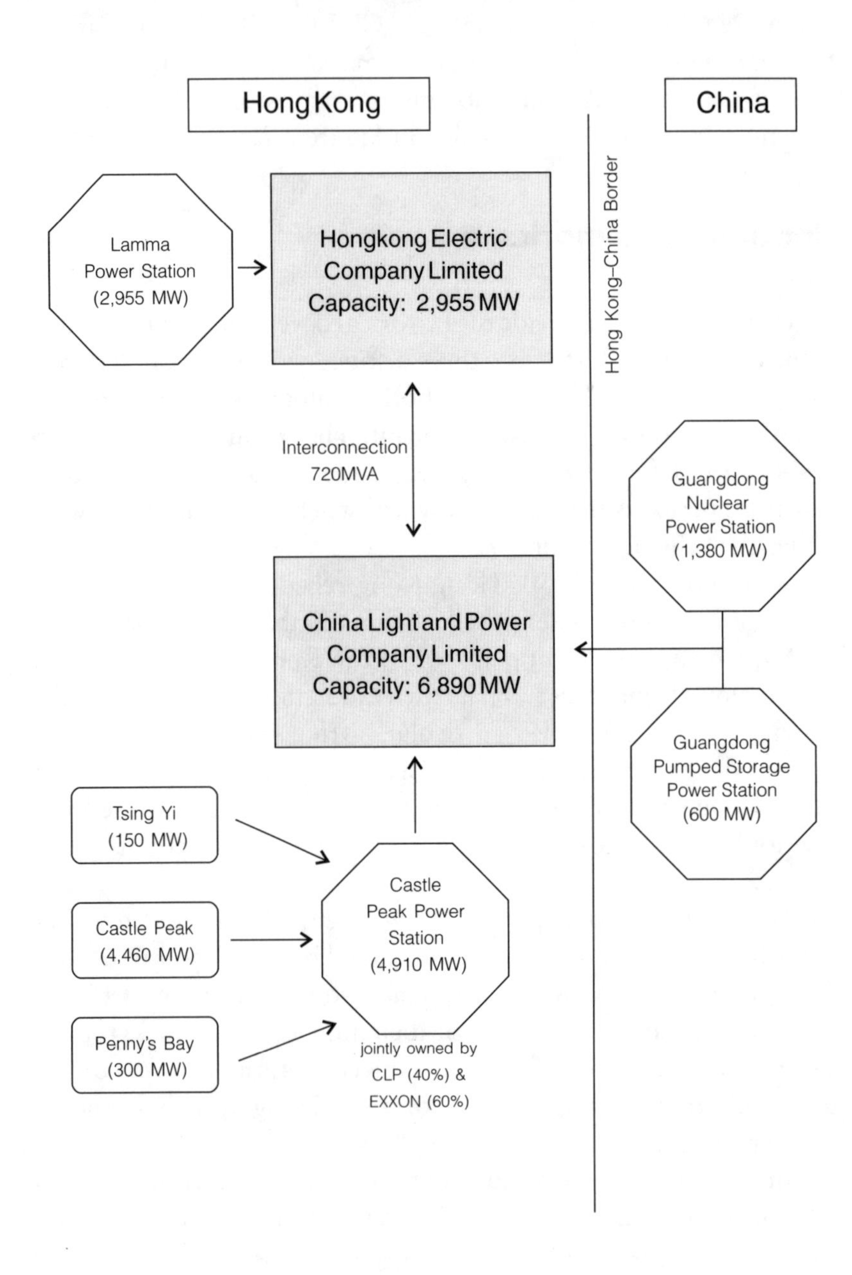

relocated from Hong Kong to China. Hong Kong transformed itself gradually from an industrial city to a service centre. This transformation process was clearly indicated by a decline in industrial demand and an increase in commercial demand for electricity.

## The Gas Industry

### The Development of Towngas before the 1980s

In Hong Kong towngas is mainly for domestic and commercial use. The Hong Kong and China Gas Company Limited (HKCG), the only supplier of towngas, was first established in 1862. The company initially contracted with the Hong Kong government to supply towngas for street lighting. During World War II the company's plants at Ma Tau Kok (in Kowloon) and West Point (on Hong Kong Island) were operated by the Japanese until stocks of coal were exhausted. The plants were then left to stand idle for the last two years of the war. They were brought into operation again in January 1946, and service to the public was gradually recovered as mains and piping were replaced. By 1950 repair work was completed and production returned to the pre-war level. These plants produced towngas from heavy fuel oil.

In the 1950s, as the demand for gas increased, a new plant was installed to replace the old one at Ma Tau Kok. Twin underwater gas mains were laid across Victoria Harbour in 1958 to connect the Kowloon and Island areas of supply. Construction of an additional plant at Ma Tau Kok was also carried out. The new plant would use oil instead of coal for gas-making. After the new oil gas plant was completed and in operation, the West Point station closed, and all gas requirements from the Hong Kong side were then pumped across the harbour from the Ma Tau Kok plant. In 1964 a new butane/air plant at Tsuen Wan commenced operations. The plant supplied towngas to factories in the rapidly expanding satellite town and operated independent of the main works. HKCG also extended its operation to another satellite town at Kwun Tong.

**Figure 2.3**
**Towngas Consumption in Hong Kong, 1952–1995**

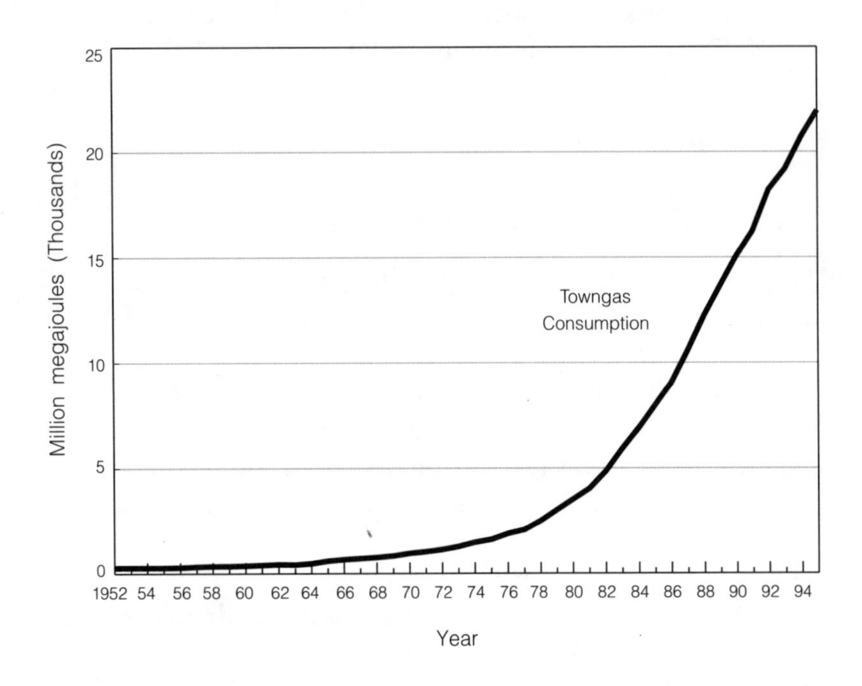


To keep production capacity ahead of the rapid increase in the demand for towngas, two new units in Ma Tau Kok were constructed and came into operation in 1972. The new units manufactured gas from naphtha, which was a new raw material at that time. Existing units were also converted for use with the new raw material.

## The Rise and Fall of LPG

Liquefied petroleum gas (LPG) is a chemical by-product of petroleum. Bottled LPG was first introduced to Hong Kong in 1956 and was mainly offered to customers who were not in the towngas supply area. As LPG was cleaner and more convenient, it soon replaced kerosene to become the major fuel used in Hong Kong. In

1980 the demand for LPG exceeded that for towngas, and LPG was the most widely used form of fuel in Hong Kong.

A turning point in the history of Hong Kong's gas industry occurred in 1981. Because of numerous accidents caused by LPG, the government invited consultants from British Gas (BG) to study the safety and legal aspects of the gas supply in Hong Kong. After the publication of the Report on the *Safety and Legal Aspects of Both Town Gas and LPG Operations in Hong Kong* (1981), the market concentration of the gas industry changed dramatically. Following the recommendations in the Report (1981), the government introduced a piped gas policy to discourage the use of bottled gas in domestic dwellings. At the same time, it encouraged the upgrading of sub-standard gas water heaters. The Gas Safety (Gas Supply) Regulations prohibited the installations of any gas mains for the conveyance of LPG along or across a road. The transmission of LPG under public roadways was banned completely. These constraints, imposed by the government in the 1980s, raised HKCG's competitive power and directly led to the company's success.

## Rapid Expansion of Towngas in the 1980s

Figure 2.3 shows the consumption of towngas in Hong Kong. In the 1950s and 1960s the business of HKCG grew very slowly. Despite this early and unsuccessful introduction of towngas, HKCG continued to invest in its production and distribution facilities, and at the same time tried to promote its product to the community. When the living standards of Hong Kong people improved in the 1970s and 1980s, the consumption of towngas increased remarkably. To meet increasing demand from the new towns in the New Territories HKCG constructed a second gas production plant in Tai Po. It began in March 1986 and was completed in 1993. In 1995 the Tai Po plant produced 96% of the total amount of gas supplied by HKCG. Further extensions of HKCG's distribution mains increased the length of its distribution system to 1,941 kilometres in 1995, bringing towngas within economic reach of an estimated 80% of Hong Kong homes.

As an interim measure intended to meet the increased demand for gas, the government allowed HKCG to provide substitute natural gas (SNG) to customers in the western part of the New Territories. SNG was produced from an LPG/air mix at temporary plants located in the new towns of Yuen Long and Tuen Mun. The gas was supplied through separate distribution systems within each township. These temporary gas works were scheduled to be decommissioned when natural gas became available and when the towngas high pressure transmission pipelines reached these two towns in the early 1990s.

In order to provide a secure, long-term supply of gas to Hong Kong as economically as possible, HKCG has planned to import natural gas from the South China Sea. The company has also considered the import of liquefied natural gas (LNG) from other countries in Southeast Asia. Despite its early attempt to import natural gas, the company lost a bid (to CLP) to purchase natural gas at a newly discovered gas field off Hainan Island.

The business of HKCG continued to grow at a very rapid rate in the 1990s. To meet future demand for towngas from the new Chek Lap Kok Airport and other developments, the company has commenced work on a project to lay a transmission pipeline from Route Twisk to Tai Ho on Lantau Island. Gas supply is scheduled to commence in December 1996. HKCG has also continued to seek investment opportunities in the gas industry in the People's Republic of China. In 1994 the company concluded joint venture contracts with the cities of Panyu and Zhongshan in the Pearl River Delta region of Guangdong Province. In 1995 a contract was signed with the Guangzhou Gas Company to build a SNG plant and to supply gas by pipeline to residents in the Fangcun district of Guangzhou.

In recent years, like the other two electricity companies, HKCG has diversified into the property development business. In 1990, HKCG decided to redevelop its North Point Depot site into a twenty-five-storey Operations Building. In 1996 the company joined a consortium to develop the Hong Kong Island terminus of the Airport Railway. The company is also considering the redevelopment of the Ma Tau Kok plant site. As Kai Tak Airport is to close upon the completion of Chek Lap Kok Airport in early 1998, subsequent environmental upgrading of the whole of Southeast Kowloon may require the removal of the plant. The company has proposed that, with appropriate assistance and compensations for removal, it would substitute the Ma Tau Kok capacity with facilities at a new site.

## Energy Requirements in Hong Kong

Since Hong Kong does not have its own indigenous resources, different forms of primary energy have to be imported. Primary energy refers to natural resources extracted from a stock of reserves in the ground (e.g., coal, crude oil, natural gas) without undergoing any transformation processes. Imported primary energy can be used for energy transformation or final consumption. In contrast, secondary energy or transformed energy is obtained from the transformation of primary energy. For example, towngas is manufactured through the transformation of imported naphtha; the generation of electricity requires heavy fuel oil or steam coal as a fuel source.

Table 2.1 shows the overall energy balance of Hong Kong from 1979 to 1995. The overall energy balance summarizes the origins and uses of all forms of energy. As is shown in Table 2.1, energy requirements can be basically classified into two types: primary energy requirements and final energy requirements. Primary energy requirements (PER) refer to all types of oil and coal products. Final energy requirements (FER), on the other hand, refer to the amount of energy consumed by final users for purposes such as heating, cooking, and motive power for driving machinery, but exclude non-energy uses such as kerosene as solvent. Energy consumed by final users can be either primary energy (that has not undergone any transformation) or secondary energy.

**Table 2.1**
**Overall Energy Balance of Hong Kong, 1979–1995**

**(1) Primary energy requirements (terajoule TJ)**

| Year | Coal products (a) | Oil products (b) | Net export of electricity (c) | Total (a)+(b)–(c) |
|---|---|---|---|---|
| 1979 | 942 | 195,369 | 906 | 195,405 |
| 1980 | 869 | 203,774 | 1,109 | 203,534 |
| 1981 | 849 | 216,931 | 851 | 216,929 |
| 1982 | 37,301 | 210,708 | 979 | 247,030 |
| 1983 | 97,218 | 171,735 | 1,323 | 267,630 |
| 1984 | 132,961 | 156,560 | 2,664 | 286,857 |
| 1985 | 158,086 | 136,020 | 3,781 | 290,326 |
| 1986 | 192,759 | 141,551 | 4,350 | 329,960 |
| 1987 | 234,470 | 126,691 | 4,904 | 356,258 |
| 1988 | 257,066 | 146,115 | 5,185 | 397,996 |
| 1989 | 273,687 | 149,426 | 6,371 | 416,742 |
| 1990 | 293,856 | 142,767 | 6,470 | 430,153 |
| 1991 | 318,689 | 160,290 | 11,020 | 467,958 |
| 1992 | 339,804 | 183,646 | 17,866 | 505,584 |
| 1993 | 350,757 | 189,703 | 16,202 | 524,257 |
| 1994 | 271,612 | 203,828 | –23,383 | 498,823 |
| 1995 | 281,311 | 192,241 | –21,824 | 495,375 |

**(2) Fuel inputs and system loss (TJ)**

| Year | Coal used for electricity generation | Oil used for electricity generation | Oil used for gas production | System loss in electricity generation |
|---|---|---|---|---|
| 1979 | 0 | 110,420 | 3,109 | 4,536 |
| 1980 | 0 | 122,928 | 3,601 | 5,110 |
| 1981 | 0 | 128,640 | 4,205 | 5,441 |
| 1982 | 36,490 | 106,933 | 5,293 | 6,475 |
| 1983 | 96,388 | 71,477 | 6,302 | 7,511 |
| 1984 | 132,186 | 51,119 | 7,601 | 7,694 |
| 1985 | 157,315 | 40,016 | 8,346 | 8,126 |
| 1986 | 192,012 | 28,124 | 10,363 | 9,141 |
| 1987 | 233,772 | 10,724 | 11,076 | 9,981 |
| 1988 | 256,475 | 5,284 | 13,383 | 11,010 |
| 1989 | 273,061 | 7,898 | 14,544 | 11,539 |
| 1990 | 293,278 | 5,544 | 16,017 | 11,905 |
| 1991 | 318,111 | 12,264 | 17,274 | 12,346 |
| 1992 | 339,255 | 16,152 | 19,369 | 13,672 |
| 1993 | 350,145 | 17,608 | 20,871 | 15,005 |
| 1994 | 271,109 | 5,469 | 23,645 | 14,595 |
| 1995 | 280,872 | 8,172 | 23,193 | 14,843 |

**(3) Final energy requirements (TJ)**

| Year | Coal products | Oil products | Electricity | Gas | Total |
|---|---|---|---|---|---|
| 1979 | 942 | 81,840 | 35,564 | 3,021 | 121,367 |
| 1980 | 869 | 77,244 | 39,318 | 3,524 | 120,956 |
| 1981 | 849 | 84,085 | 41,620 | 4,034 | 130,588 |
| 1982 | 811 | 98,482 | 44,769 | 4,858 | 148,920 |
| 1983 | 830 | 93,957 | 50,526 | 5,924 | 151,237 |
| 1984 | 775 | 97,841 | 54,166 | 6,907 | 159,689 |
| 1985 | 771 | 87,658 | 57,340 | 7,979 | 153,749 |
| 1986 | 746 | 103,064 | 63,592 | 9,043 | 176,445 |
| 1987 | 698 | 104,891 | 70,626 | 10,584 | 186,799 |
| 1988 | 592 | 127,448 | 75,633 | 12,247 | 215,919 |
| 1989 | 626 | 126,984 | 80,589 | 13,671 | 221,870 |
| 1990 | 577 | 121,205 | 85,801 | 15,056 | 222,640 |
| 1991 | 578 | 130,752 | 91,140 | 16,238 | 238,707 |
| 1992 | 549 | 148,125 | 94,151 | 18,207 | 261,032 |
| 1993 | 611 | 151,224 | 99,810 | 19,198 | 270,843 |
| 1994 | 503 | 174,714 | 105,055 | 20,727 | 300,999 |
| 1995 | 439 | 160,876 | 107,477 | 21,972 | 290,764 |

Source: *Hong Kong Energy Statistics*.

# Pattern of Energy Consumption

## Consumption before the Early 1980s

Before the 1950s energy was mostly derived from solid fuels, such as firewood, charcoal, coke, and coal. In 1950, 35 per cent of Hong Kong's energy was derived from solid fuels (Chou 1979). But a transition was underway thereafter towards liquid fuels like oil and LPG. Liquid fuels were preferred for importation over solid fuels because they are less bulky and more convenient to handle. By the early 1960s kerosene replaced firewood and charcoal to become the main domestic fuel in Hong Kong. From the time of its introduction in the late 1950s, the consumption of LPG grew rapidly and became the main domestic fuel in Hong Kong during the 1970s.

Compared with other newly industrialized countries in the region such as Taiwan, Singapore, and South Korea, Hong Kong has historically had a relatively low energy intensity, or energy-

**Table 2.2**
**Local Energy Demand by Sector, 1970–1995**
**(1) Electricity consumption (terajoule TJ)**

| Year | Domestic | Commercial | Industrial | Total |
|---|---|---|---|---|
| 1970 | 3,338 (21) | 6,096 (38) | 6,588 (41) | 16,023 |
| 1971 | 3,812 (22) | 6,484 (37) | 7,312 (41) | 17,609 |
| 1972 | 4,296 (22) | 7,227 (37) | 7,961 (41) | 19,484 |
| 1973 | 4,760 (22) | 8,166 (38) | 8,712 (40) | 21,638 |
| 1974 | 5,017 (24) | 7,935 (37) | 8,349 (39) | 21,302 |
| 1975 | 5,507 (24) | 8,775 (38) | 8,843 (38) | 23,126 |
| 1976 | 5,940 (23) | 9,716 (37) | 10,535 (40) | 26,190 |
| 1977 | 6,944 (23) | 11,348 (38) | 11,499 (39) | 29,792 |
| 1978 | 7,364 (22) | 12,650 (39) | 12,781 (39) | 32,794 |
| 1979 | 7,556 (21) | 13,883 (39) | 14,124 (40) | 35,563 |
| 1980 | 8,414 (21) | 15,779 (40) | 15,125 (39) | 39,318 |
| 1981 | 8,733 (21) | 17,481 (42) | 15,406 (37) | 41,620 |
| 1982 | 9,227 (21) | 20,106 (45) | 15,436 (34) | 44,769 |
| 1983 | 10,685 (21) | 23,021 (46) | 16,820 (33) | 50,526 |
| 1984 | 10,817 (20) | 24,806 (46) | 18,543 (34) | 54,166 |
| 1985 | 11,519 (20) | 27,001 (47) | 18,820 (33) | 57,340 |
| 1986 | 12,808 (20) | 29,393 (46) | 21,391 (34) | 63,592 |
| 1987 | 14,022 (20) | 32,625 (46) | 23,979 (34) | 70,626 |
| 1988 | 15,712 (21) | 35,045 (46) | 24,876 (33) | 75,633 |
| 1989 | 17,075 (21) | 38,336 (48) | 25,178 (31) | 80,589 |
| 1990 | 19,037 (22) | 41,830 (49) | 24,934 (29) | 85,801 |
| 1991 | 20,586 (23) | 45,503 (50) | 25,051 (27) | 91,140 |
| 1992 | 21,716 (23) | 48,242 (51) | 24,194 (26) | 94,151 |
| 1993 | 24,092 (24) | 52,485 (53) | 23,233 (23) | 99,810 |
| 1994 | 25,827 (25) | 57,790 (55) | 21,437 (20) | 105,055 |
| 1995 | 27,063 (25) | 60,190 (56) | 20,222 (19) | 107,476 |

Note: Figures in brackets are percentage shares.

income ratio. Income elasticity of energy demand is also relatively low. Several factors account for the low energy intensity and income elasticity. First, Hong Kong does not specialize in heavy industries, such as iron and steel, chemicals, and fertilizers, which are regarded as highly energy-intensive industries. Light industries that have developed in Hong Kong are not energy-intensive. Primary energy in Hong Kong is chiefly used for energy transformation rather than for industrial production. Second, the compact urban society feature of Hong Kong and the strict government controls on the number of motor vehicles have reduced the consumption of energy in the transport sector. Third, the energy

**(2) Towngas consumption (TJ)**

| Year | Domestic | Commercial | Industrial | Total |
|---|---|---|---|---|
| 1970 | 494 (51) | 357 (37) | 107 (12) | 958 |
| 1971 | 543 (53) | 392 (37) | 100 (10) | 1,035 |
| 1972 | 587 (53) | 441 (38) | 107 ( 9) | 1,135 |
| 1973 | 693 (54) | 480 (38) | 104 ( 8) | 1,278 |
| 1974 | 790 (54) | 582 (39) | 104 ( 7) | 1,476 |
| 1975 | 850 (53) | 639 (40) | 120 ( 7) | 1,609 |
| 1976 | 930 (49) | 801 (42) | 160 ( 9) | 1,891 |
| 1977 | 994 (48) | 904 (43) | 180 ( 9) | 2,077 |
| 1978 | 1,249 (50) | 1,076 (43) | 179 ( 7) | 2,503 |
| 1979 | 1,501 (50) | 1,318 (43) | 202 ( 7) | 3,021 |
| 1980 | 1,758 (50) | 1,554 (44) | 212 ( 6) | 3,524 |
| 1981 | 2,025 (50) | 1,784 (44) | 225 ( 6) | 4,034 |
| 1982 | 2,500 (51) | 2,138 (44) | 220 ( 5) | 4,858 |
| 1983 | 3,021 (51) | 2,670 (45) | 233 ( 4) | 5,924 |
| 1984 | 3,476 (50) | 3,174 (46) | 257 ( 4) | 6,907 |
| 1985 | 4,036 (51) | 3,669 (46) | 274 ( 3) | 7,979 |
| 1986 | 4,593 (51) | 4,123 (45) | 327 ( 4) | 9,043 |
| 1987 | 5,254 (50) | 4,930 (46) | 399 ( 4) | 10,584 |
| 1988 | 6,127 (50) | 5,680 (46) | 440 ( 4) | 12,247 |
| 1989 | 6,943 (51) | 6,218 (45) | 510 ( 4) | 13,671 |
| 1990 | 7,596 (50) | 6,877 (46) | 583 ( 4) | 15,056 |
| 1991 | 8,133 (50) | 7,404 (46) | 701 ( 4) | 16,238 |
| 1992 | 9,152 (50) | 8,232 (45) | 823 ( 5) | 18,207 |
| 1993 | 9,657 (50) | 8,652 (45) | 889 ( 5) | 19,198 |
| 1994 | 10,606 (51) | 9,202 (44) | 919 ( 4) | 20,727 |
| 1995 | 11,408 (52) | 9,586 (44) | 978 ( 4) | 21,972 |

Note: Figures in brackets are percentage shares.
Source: *Hong Kong Energy Statistics*.

requirements for heating and cooling are relatively low in Hong Kong, as Hong Kong does not have a real need for winter heating.

As is shown in Table 2.1, most of the primary energy in Hong Kong was used to generate two main kinds of power, electricity and gas. Electricity has been more popular than towngas. Before 1982 the most important oil product imported into Hong Kong was fuel oil used for electricity generation. In Hong Kong, more than half of the primary energy requirements is consumed in the generation of electricity, with the commercial sector being the largest user of electricity (see Table 2.2). Apart from electricity generation, primary energy is mostly required by the transport and the

**Table 2.3 Retained Imports of Oil and Coal Products, 1979–1995**

**(1) Retained imports of oil products**

| Year | Aviation gasoline and aviation kerosene (kilolitre) | Motor gasoline (kilolitre) | | Kerosene (kilolitre) | Gas oil, diesel oil and naphtha | Fuel oil | LPG (tonne) |
|---|---|---|---|---|---|---|---|
| 1979 | 746,497 | 243,638 | | 300,759 | 1,204,966 | 3,898,216 | 105,089 |
| 1980 | 744,260 | 257,759 | | 306,605 | 1,099,090 | 4,059,277 | 103,350 |
| 1981 | 950,105 | 343,471 | | 208,947 | 1,171,816 | 4,245,461 | 113,969 |
| 1982 | 1,042,359 | 320,492 | | 120,205 | 1,405,126 | 3,952,571 | 126,852 |
| 1983 | 1,023,414 | 279,799 | | 82,326 | 1,406,525 | 2,943,288 | 130,151 |
| 1984 | 1,090,596 | 269,309 | | 207,101 | 1,396,251 | 2,511,379 | 144,847 |
| 1985 | 1,195,676 | 269,087 | | 82,459 | 1,258,087 | 2,056,152 | 143,189 |
| 1986 | 1,420,028 | 246,431 | | 89,636 | 1,734,095 | 2,187,464 | 161,794 |
| 1987 | 1,470,219 | 266,082 | | 117,191 | 1,703,894 | 1,866,357 | 163,988 |
| 1988 | 1,690,850 | 264,919 | | 74,502 | 1,801,863 | 1,671,389 | 159,533 |
| 1989 | 2,000,557 | 311,871 | | 31,972 | 1,890,198 | 1,988,795 | 176,783 |
| 1990 | 2,296,055 | 388,524 | | 7,207 | 2,690,712 | 1,347,198 | 169,224 |
| 1991 | 2,105,704 | 350,968 | | 3,756 | 2,889,706 | 1,152,023 | 163,379 |
| | | Leaded petrol | Unleaded petrol | | | | |
| 1992 | 2,690,530 | 221,549 | 277,129 | –10,958 | 3,743,913 | 1,881,956 | 194,865 |
| 1993 | 2,693,652 | 128,775 | 238,350 | –10,927 | 3,638,106 | 1,552,543 | 130,685 |
| 1994 | 3,438,787 | 144,872 | 348,899 | –7,154 | 4,823,238 | 1,580,435 | 105,666 |
| 1995 | 3,318,386 | 188,596 | 323,351 | N.A. | 4,691,324 | 1,561,524 | 123,705 |

**(2) Retained imports of coal products (tonne)**

| Year | Steam coal and other coal | Wood charcoal | Anthracite | Coke and semi-coke | Total |
|---|---|---|---|---|---|
| 1979 | 509 | 19,634 | 5,305 | 6,746 | 32,194 |
| 1980 | 756 | 21,406 | 1,995 | 5,561 | 29,718 |
| 1981 | 51,696 | 19,927 | 5,304 | 3,735 | 80,662 |
| 1982 | 1,451,485 | 21,959 | 3,191 | 2,520 | 1,479,155 |
| 1983 | 3,416,006 | 24,760 | 2,307 | 1,253 | 3,444,326 |
| 1984 | 4 ,459,637 | 21,359 | 2,598 | 2,489 | 4,486,083 |
| 1985 | 5,520,669 | 19,852 | 2,515 | 3,951 | 5,546,987 |
| 1986 | 6,390,253 | 19,844 | 2,265 | 3,360 | 6,415,722 |
| 1987 | 8,007,311 | 18,646 | 2,259 | 2,909 | 8,031,125 |
| 1988 | 9,264,886 | 16,697 | 1,576 | 1,917 | 9,285,076 |
| 1989 | 9,925,879 | 18,873 | 1,960 | 517 | 9,947,229 |
| 1990 | 8,928,614 | 16,252 | 2,053 | 1,404 | 8,948,323 |
| 1991 | 9,633,042 | 17,221 | 2,092 | 402 | 9,652,757 |
| 1992 | 10,213,195 | 15,958 | 1,707 | 1,083 | 10,231,943 |
| 1993 | 11,828,369 | 16,828 | 1,290 | 2,750 | 11,849,237 |
| 1994 | 8,450,362 | 14,711 | 550 | 1,893 | 8,467,516 |
| 1995 | 9,108,994 | 13,920 | 0 | 1,063 | 9,123,977 |

Source: *Hong Kong Energy Statistics.*

Note: Starting from 1995, kerosene and aviation kerosene are classified under the same item. The negative value was due to the fact that some of the imported aviation kerosene was regarding to kerosene for re-export.

industrial sectors. The common oil products used by the transport sector include aviation fuels for planes, motor gasoline, automobile diesel and diesel fuels. The industrial sector uses mainly industrial diesel oil and other heavy oils for production (see Table 2.3).

## Consumption since the Early 1980s

A significant change occurred in 1982 in the primary energy requirements. The introduction of coal-fired plants in electricity generation in that year caused a dramatic surge in the retained imports of steam coal. Since then, the percentage share of coal fuels in electricity generation has increased, while that of oil fuel has decreased significantly. Another major change in primary energy requirements since the early 1980s is the sharp increase in the import of naphtha used for towngas production. As a result of the government's safety policy and HKCG's promotion effort, the consumption of towngas has increased substantially since 1981. This increase in the import of naphtha, coupled and with the use of coal in electricity generation, has caused the share of primary energy used for transformation to increase steadily from 58 per cent in 1979 to 63 per cent in 1995. LPG imports, on the other hand, increased only moderately until 1992, after which they decreased sharply (see Table 2.3).

Despite the substantial changes that have occurred in the proportion of coal fuels and oil fuels used in power generation since the early 1980s, these fuels are still primarily required for the generation of electricity and gas. Power generation (electricity and gas) remains the major use of the imported primary energy in Hong Kong. As a result of structural change of the Hong Kong economy, however, the energy demand for electricity by sector has changed substantially over time (see Table 2.2).

In its process of industrialization in the 1960s and 1970s, Hong Kong concentrated on the development of light industries such as the textiles and clothing, plastics and electronics industries. The demand for motive power in these light industries exceeded that for heat energy. Since the late 1970s Hong Kong has been transforming

**Table 2.4**
**Number of Households Piped for Central Gas Supply, 1988–92**

| | Private housing | | Housing Authority | | Housing Society | | Total market | |
|---|---|---|---|---|---|---|---|---|
| Year | HKCG | LPG | HKCG | LPG | HKCG | LPG | HKCG (%) | LPG (%) |
| 1988 | 370,518 | 48,318 | 421,676 | 127,003 | 18,511 | 9,537 | 810,705 (81) | 184,858 (19) |
| 1989 | 401,049 | 49,546 | 454,279 | 128,437 | 18,908 | 11,260 | 874,236 (82) | 189,243 (18) |
| 1990 | 423,920 | 50,697 | 501,884 | 132,171 | 19,417 | 11,952 | 945,221 (83) | 194,820 (17) |
| 1991 | 456,920 | 53,560 | 544,486 | 128,779 | 20,831 | 15,388 | 1,022,237 (84) | 197,727 (16) |
| 1992 | 483,522 | 56,360 | 567,395 | 128,779 | 21,107 | 15,388 | 1,072,024 (84) | 200,527 (16) |

Source: HKCG; Consumer Council (1995).

**Table 2.5**
**Overall Efficiency of Energy Delivery**

| | Town gas | LPG | | Electricity (coal/gas) |
|---|---|---|---|---|
| Gas/oil well | 96.9% | 96.9% | Mining/extraction | 99.2%/99.6% |
| Tanker (to refinery) | 98.8% | 98.5% | | |
| Refinery | 93.7% | 93.7% | | |
| Tanker (to Hong Kong) | 99.5% | 99.6% | Ship/Pipeline | 97.9%/96.4% |
| Gas production plant | 90.3% | | Electricity plant | 34.1%/52.0% |
| Distribution | 96.5% | 95.0% | Distribution | 95.3%/95.3% |
| Water heater (instantaneous) | 79.0% | 79.0% | Water heater (instantaneous) | 97.0%/97.0% |
| Overall efficiency | 61.4% | 66.9% | Overall | 30.6%/44.7% |

Source: *Asia Engineer* (November 1995).

**Table 2.6**
**Estimated Emissions of Delivered Hot Water**

| | Emission kg/Tj | | | | |
|---|---|---|---|---|---|
| Fuel | $SO_2$ | $NO_x$ | $CO_2$ | Particulate | Solid waste |
| Electricity (coal) | 1,330 | 1,090 | 282,520 | 70 | 8,240 |
| Electricity (coal w/FGD) | 210 | 1,090 | 282,520 | 70 | 16,400 |
| Natural gas | 20 | 280 | 110,770 | 0 | 10 |
| Town gas | 80 | 170 | 107,010 | 0 | 90 |
| LPG | 80 | 180 | 87,110 | 0 | 80 |

Source: *Asia Engineer* (November 1995).

itself from an industrial city to a service centre. Hong Kong is now one of the major financial centres within the region. Banking industries, insurance business, and other commercial sector activities have been growing rapidly. Relocation of factories to China, further led to the rise in the demand for electricity in the commercial sector relative to that in industrial sector. At present, the commercial sector accounts for the greatest level of consumption of the electricity generated.

With respect to the transport sector, the energy demand for electricity has not been important. This is because most of the modes of transportation consume oil products directly. However, with the rapid development of electrified modes of transportation like the electrified Kowloon–Canton Rail Transit, the Mass Transit Railway, and the Light Railway, there has been a small increase in the demand for electricity in the transport sector since the 1980s.

Unlike electricity, which is mainly consumed in the production of motive power, gas is used primarily for producing heat energy. Gas is especially suitable for domestic use (cooking and heating), and also for commercial use in restaurants and hotels. As can be seen in Table 2.2, domestic consumption accounted for more than 50 per cent of total towngas consumption in 1995. Although the usage of gas is rather limited, towngas has experienced the fastest growth rate in recent years. As more and more new housing units are fitted with pipes for the supply of towngas, the market share of towngas in the fuel market is expected to grow in the future (see Table 2.4).

As the income of an economy rises, there is an upward transition in the selection of household fuels (Chow 1989). People tend to upgrade the type of fuel used when their income increases. They start to shift to fuels that are cleaner, healthier, more convenient, and easier to handle. In the 1960s solid fuels were replaced by liquid fuels, while in the 1980s liquid fuels were replaced by gas fuels. As towngas is of higher quality in terms of cleanliness and ease of handling, most families nowadays prefer to cook with towngas instead of other solid or liquid fuels. In addition, the demand for consumer durables increases with a rise in income, and this in turn raises the residential demand for electricity in Hong Kong.

## Energy Efficiency and Conservation

Fuels used in electricity generation have their respective drawbacks. Oil-fired generation is risky in terms of oil price fluctuations and import dependence. Besides, there are environmental effects associated with its use. Although the use of coal fuels for electricity generation can lower fuel costs, it too has problems. Apart from a higher capital expenditure incurred in constructing coal-fired generators, generating electricity with coal fuels can cause serious environmental problems and lower the thermal efficiency in energy transformation. The drawbacks include acid rain, greenhouse gas emissions, toxic air, and solid wastes.

When energy is transformed from primary energy to secondary energy, there is always some conversion loss. Conversion loss refers to energy loss during the transformation and distribution processes in energy production. As is shown in the overall energy balance in Table 2.1, the growth of the primary energy requirements has far exceeded that of final energy requirements since the early 1980s. The slower growth rate of final energy requirements is due to the reduction in thermal efficiency of the local energy sector. Compared with the use of oil fuels, steam coal used in electricity generation has reduced local thermal efficiency due to a relatively larger conversion loss in energy transformation. According to the results of a study (Hutchison et al. 1995), overall energy efficiency of electricity production in Hong Kong is much lower when compared with towngas or LPG production (Table 2.5).

In addition to greater energy loss, electricity generation produces a higher level of air pollution in the energy transformation process. Table 2.6 shows the estimated emission levels of various types of energy. Despite the fact that the two electricity companies have planned to lower their air pollution emissions through various controls and installations, these measures are not expected to reduce emission to levels comparable to those resulting from gas production. The Hutchison study therefore concludes that towngas and LPG are more efficient and environmentally friendly as compared with electricity generated by coal fuel. Nevertheless, the study points out the fact that electricity is more versatile and has a much wider range of applications than gas does.

As the income level in Hong Kong increases, people are becoming more concerned about the quality of living. The Environmental Protection Department (EPD), which was formally established in April 1986, charged the responsibilities for environmental protection with some government departments. The Department designs policies on controlling air, water, and noise pollution in Hong Kong. When CLP designed its new power station at Black Point, the company decided that the new units would be fired by natural gas, which could satisfy the standards set by the EPD. With the introduction of natural gas supply for Black Point, CLP's base and intermediate power generation will be split evenly among coal, natural gas, and nuclear power.

In comparison to oil and coal, the environmental performance of natural gas favours its use in three key ways. First, the costs of meeting air pollution standards for sulphur dioxide and nitrogen are generally lowest for natural gas. Second, technological advances have raised the thermal efficiency of natural gas-fired units to 50 per cent. Third, the planning horizon and the lead time required for the construction of gas-fired units are shorter. This allows greater flexibility in adjusting commissioning dates when demand changes unexpectedly.

In addition to lowering emissions from generating plants, EPD also to reduce vehicular emissions to improve the air quality in Hong Kong. In 1992 unleaded petrol was introduced in Hong Kong (see Table 2.3). The government has provided incentives in terms of a lower fuel tax to encourage the use of unleaded petrol. Consequently, the relative market share of leaded petrol has decreased since 1992. The government also has a policy to convert all diesel-powered vehicles to run on petrol as a means of reducing vehicular emissions.

As two-thirds of primary energy in Hong Kong is used for the production of electricity and towngas, and LPG is still used as an important domestic fuel in Hong Kong, the discussion of government policies on the energy industry in subsequent chapters will be restricted to the electricity and gas industries. Fuel policies on the transport sector will not be dealt with in our discussion. Readers who are interested in learning more about the government's fuel policy on the transport sector and vehicular emissions policy in Hong Kong can refer to Hall (1996) and Rusco and Walls (1995).

# CHAPTER 3

# Government Policies on the Energy Industry

## Introduction

Over the past decade, Hong Kong has been undergoing a process of democratization in its political system. We have seen intensified public debates over the quality and pricing of utility services. Proposals of energy price increases have inevitably been followed by political opposition. Certain arguments against price increases were found to be logically inconsistent. For example, some members of the Legislative Council have argued for the removal of the Scheme of Control which protects the returns of the two electricity companies, but these Legco members have also championed a profit control scheme which would be imposed on the unregulated towngas company so as to limit price increases. Should a profit control scheme be imposed to protect customers?

In this chapter we first explain the existing government policies on the electricity and gas companies. We then analyze the problems with the existing monitoring arrangements and identify those areas that may be inhibiting free and fair competition in the energy industry. The last section of this chapter compares the performance of utilities within the power industry in Hong Kong.

## Monitoring Arrangements on Electricity Companies

In the early 1980s the two electricity companies increased tariffs sharply as a result of rising fuel prices and the construction of new coal-fired generators. The tariff increases caused a great deal of public resentment. A call for public monitoring ensued. In 1982, amid the protests against higher tariffs, the government appointed Burns & Roe Inc., consultants to review the technical aspects of system planning by the two companies. After almost a year of study, the consultants submitted their report in mid 1983. An executive summary of the report was subsequently made available to the public. The report confirmed that plans made by the two companies had been soundly based and properly applied, and that their generation development plans would provide a reliable and adequate supply without over-expansion.

In 1983 the government appointed independent consultants, Ernst & Whinney, to assess the government's arrangement for monitoring the two electricity companies and to recommend means to enhance the government's monitoring capabilities. A consultancy report was published in March 1985. The consultants concluded that the government's monitoring system in the past had been adequate and appropriate to ensure compliance with the terms of the Scheme of Control (*Hong Kong Annual Report* 1986 and 1989). Nevertheless, the consultants recommended a number of steps to improve the monitoring process. The recommendations were later considered by a special working party responsible to the Secretary for Economic Services. One recommendation, immediately adopted by the government, was that the consultants be retained to examine the technical aspects of future financing plans submitted by the two electricity companies. The working party submitted its report to the Executive Council in 1987, and the consultants' recommendations have since been implemented (*Hong Kong Annual Report* 1989).

As shown in the consultancy report (Ernst & Whinney 1984), there is no particular commission or committee responsible for the monitoring work. Several government organizations share the monitoring responsibilities. They are:

1. Treasury Financial Monitoring Unit
   Headed by a Chief Treasury Accountant, the Unit is

responsible for the review of policy regarding financial monitoring of public utilities; it is also responsible for giving financial advice regarding public utilities to the Secretary for Economic Services.

2. Economic Services Branch
   The branch is headed by the Secretary for Economic Services and is responsible for developing and recommending government economic policies including energy and electricity matters. The Secretary is accountable to the Financial Secretary. The branch has two divisions: Services and Analysis. The energy subdivision of the Economic Services Division receives professional assistance on the economic and financial side from the Economic Analysis Division and Financial Monitoring Unit, and then makes recommendations to the Secretary for Economic Services, and ultimately to the Executive Council.

3. Electrical and Mechanical Services Department
   This department is responsible for providing technical input on financial plans and conducting major reviews of power company expansion plans. The department responds to the Economic Services Branch for policy guidance.

While confirming that the existing monitoring structure was adequate and appropriate, the consultants recommended that there should be written procedures for precise specifications of monitoring responsibilities. They recommended that the Economic Services Division should continue to co-ordinate the overall monitoring effort. Since its inception in April 1986, the Environmental Protection Department has also taken up the responsibility of monitoring emissions resulting from power generation. This division of labour in monitoring continues today.

## The Scheme of Control on Electricity Companies

### The Regulatory Mechanism

The two regional monopolists (CLP and HEC) in the electricity industry are regulated by the Scheme of Control, which was first proposed by CLP in 1964. The objectives of the scheme are to allow the regulated companies and their shareholders "to earn a return which is reasonable in relation to the risks involved and the capital invested in and retained in their business, and in return, the government has to be assured that service to the consuming public continues to be adequate to meet demand, to be efficient and of high quality, and is provided at the lowest cost which is reasonable in the light of financial and other considerations." (*The Schemes of Control Agreement* 1992)

These objectives are to be achieved by setting up a development fund. The main purpose of the fund, as stated in the scheme, is to assist in financing the acquisition of fixed assets. In addition, any difference between the (actual) profit after taxation and the permitted return will be transferred to or from the development fund (Figure 3.1). In other words, the regulated firm is not allowed to keep any excess profits above the permitted levels. Any excess profits are to be put into the fund. The fund accumulated would be used for acquiring fixed assets, and interest accrued from the fund would be transferred to a rate reduction reserve used for lowering tariffs. Thus, the development fund serves to finance expansion and stabilize the regulated firm's returns and prices.

The regulated firm and the government agreed that the development fund would constitute a liability, not an asset, of the company. At present, the permitted rates of return on equity capital and debt (including development fund) capital are fixed at 15% and 13.5%, respectively. From the permitted return, the following interest deductions have to be made in order to obtain the figure of net return:

*minus* interest payable on long-term financing up to a maximum of 8% per annum; and

**Figure 3.1**
**The Scheme of Control Mechanism of Energy Utilities, Hong Kong, 1996**

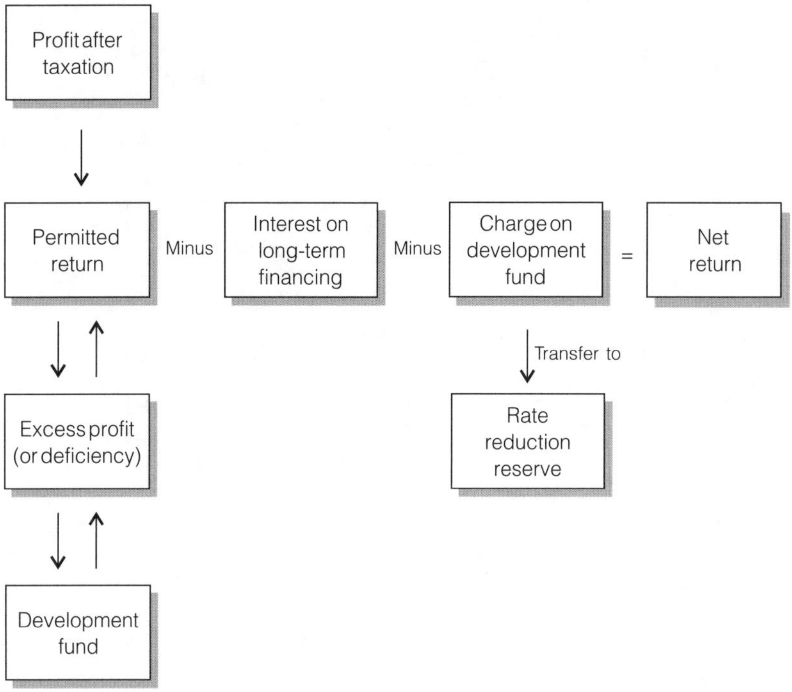

*minus* a charge of 8% per annum on the average balance of the development fund, to be credited to the rate reduction reserve, the purpose of which is to give rebates on electricity charges to consumers.

The net return is then distributed to shareholders of the company or retained for further investment. By internal transfers through the development fund, the two electricity companies were able to earn the permitted returns throughout the period under the Scheme of Control.

When either of the two electricity companies plans for major additions to its generation, transmission, and distribution system, it

**Figure 3.2**
**The Regulatory Process under the Scheme of Control**

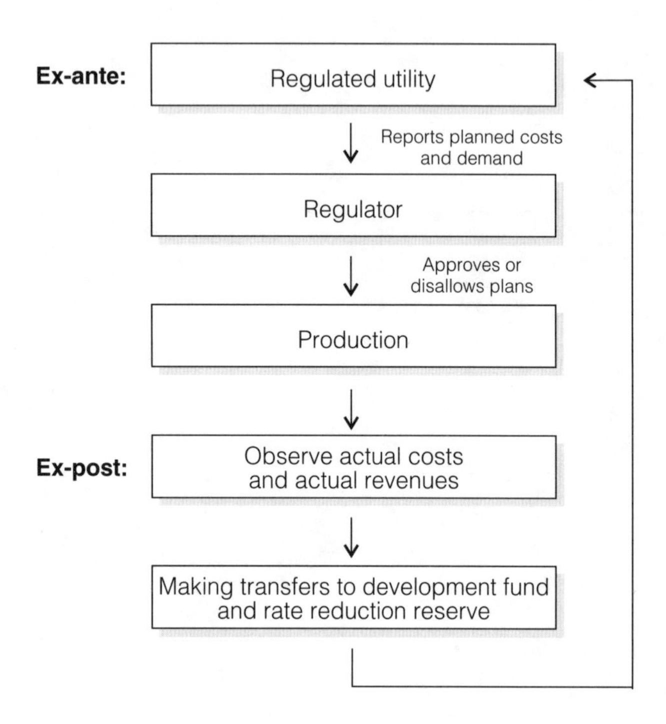


has to submit a five-year financing review to the government. The company makes projections on future demand and the basic tariffs required to meet the capital expansion. The results of each financing review is then put to the Executive Council for final approval.

In November and (or) December of each year, an audit and tariff review is conducted jointly by the government and the two electricity companies. The companies provide information to enable the government to make comparisons between actual results and the projections made in the previous financing or auditing review. They are free to adjust the basic tariff upwards if the increase does not exceed 7% of the projected tariff recently granted by the Executive Council. If the increase is in excess of 7%, the government's approval for tariff increase should be sought. Thus, built into the Scheme of Control are a number of automatic adjustment mechanisms on tariffs and returns. These mechanisms save on the cost of governing the system. In addition, the setting up of a development fund for making internal transfers helps stabilize returns and prices. Figure 3.2 shows the regulatory process under the Scheme of Control.

## Regulatory Effects of the Scheme of Control

The Scheme of Control is a formal, long-term regulatory contract designed to protect both the producer and the consumer. One of the advantages of a rate-of-return regulation over a government licensing contract is that it is easier to adjust in the event of unforeseen circumstances (Crocker and Masten 1996). But unlike the rate-of-return regulation in the U.S., which allows regulated firms to file rate cases, the Scheme of Control does not have similar flexibility in adjusting to changes in technology or demand. Once the regulator and the regulated firm have entered into a Scheme of Control Agreement, permitted rates imposed on assets are fixed for fifteen years, and the firm is allowed to raise tariffs to recover the costs of their capital investment. Although such a long-term regulatory contract can protect the regulated firm's investment against opportunistic behaviour of the regulator, it may fail to achieve efficient production decisions and may lower the incentive to reduce costs. In this section, we shall review the effects of the Scheme of Control on electricity prices and returns earned by the two electricity companies.

### *On Tariffs*

When CLP signed the Scheme of Control Agreement with the government in 1964, the company promised to reduce its tariffs in the following years. Consumers also benefited from discounts or rebates resulting from the setting up of the Rate Reduction Reserve. Hence, it was expected that the Scheme of Control would reduce the average price charged by CLP, at least for the first few years after the signing of the scheme. However, rapid increases in production costs as a result of the oil crisis caused an increase in basic tariffs in April 1974 for the first time since CLP had been subject to government control. Basic tariffs remained more or less the same for the period from 1975 to 1980.

In the early 1980s, in order to lessen Hong Kong's dependence on oil, the two electricity companies engaged in the large-scale construction of dual generators (or coal-fired generators) which could fire both oil and coal. Substantial capital expenditure was required to meet the construction costs. This, coupled with rapid oil price increases, caused the two companies to increase tariffs sharply in the late 1970s and early 1980s. From 1987 to 1991 the basic tariffs charged by CLP remained more or less unchanged, but HEC continued to increase basic tariffs every year. Although the use of coal in electricity generation can reduce fuel cost, additional capital expenditure is required to build coal-fired generators. The overall impact of building coal-fired generators on production costs and the price of electricity is not perfectly clear.

Table 3.1* (see Appendix, page 110) shows the nominal average prices of electricity charged by CLP and HEC from 1948 to 1995. We can divide the tariff history of the two companies into three periods: (1) 1948–1963, (2) 1964–1978, and (3) 1979–1995. From 1948 to 1963 both companies' prices decreased steadily. The downward trend in tariffs continued during the early years of the second period. In 1974 both companies raised prices drastically as a result of the first oil crisis. Prices then remained stable from 1975 to 1978.

CLP renewed its Scheme of Control Agreement with the

---

* All tables in Chapter 3 are given in the Appendix starting on page 110.

government in the latter part of 1978, and HEC formally entered the scheme in 1979. HEC subsequently raised its tariff significantly. The increases in tariff were partly due to the second oil crisis and partly due to the rapid capital expenditure required to finance the construction of coal-fired generators. CLP also raised prices during the same period, but to a lesser degree. After these large increases, average price remained relatively stable from 1982 to 1989. HEC has generally been charging a price that is higher than that of CLP. This may be attributed to its smaller economies of scale as compared with CLP.

From the above discussion, it can be ascertained that the first Scheme of Control (1964–1978) imposed on CLP had a greater impact on reducing tariffs. The impact on tariffs of the second Scheme of Control (1979–1993), which was imposed on both CLP and HEC, is not so clear. On the one hand, the shift from burning oil to burning coal has reduced the fuel cost of electricity; on the other hand, the government has allowed the two companies to raise basic tariffs to meet the additional capital expenditure. The overall impact is then an empirical question.

In Table 3.2 we attempt to measure the change in electricity prices in real terms. By comparing the increases in electricity prices with general price increases (consumer price index, CPI) during the same period, we are able to determine whether the real prices of electricity in Hong Kong have decreased or not. As we can see from Figures 3.3 and 3.4, the increases in electricity prices for the period 1973–1995 are more or less in line with the inflation rates over the period. In other words, real electricity prices in Hong Kong did not fall over the period.

When comparing energy prices, many people would use 1973 as the base year. After the first oil crisis, quite a few countries adopted policies to reduce their reliance on fuel oil. Consequently, the use of coal for electricity generation received a great boost following the 1973 oil price shock. Most utilities in other countries turned to coal to provide the bulk of their new base-load capacity. Some oil-fired plants were also converted to burn coal. However, a

**Figure 3.3**
**Inflation Rates and Electricity Prices of China Light and Power Company Limited, 1973–1995**

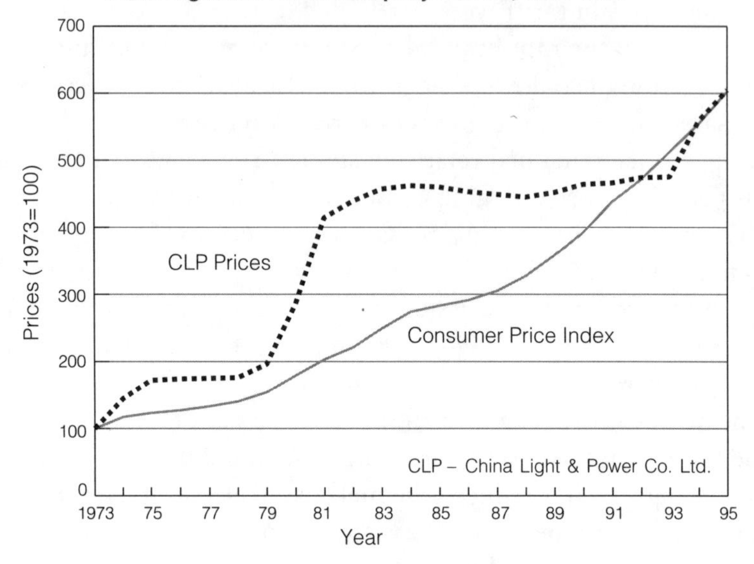


**Figure 3.4**
**Inflation Rates and Electricity Prices of Hongkong Electric Company Limited, 1973–1995**

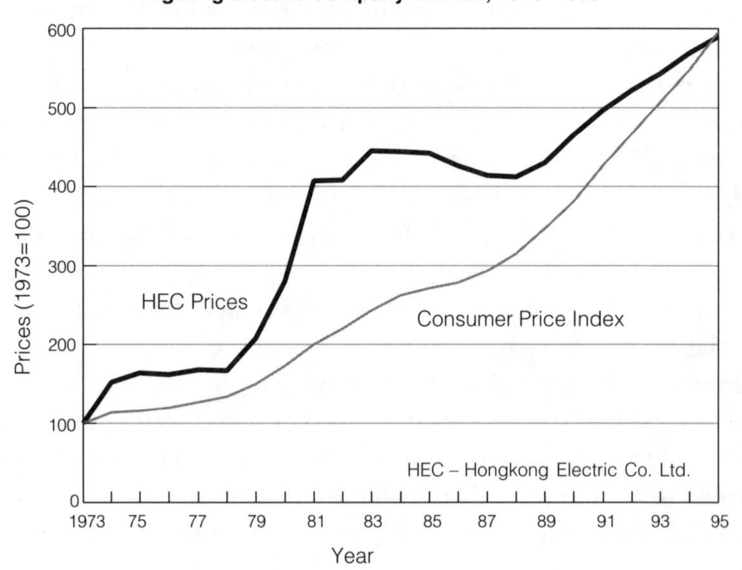

dramatic fall in oil prices since 1986 has reduced the cost effectiveness of coal-fired units over oil-fired units. This, together with the decreasing demand for electricity, has meant that coal-fired plants have turned out to be poor investments in Organisation for Economic Cooperation and Development (OECD (IEA 1992)). In Hong Kong, the two electricity companies also decided to build coal-fired units to meet demand in the 1980s. As we have shown, consumers did not really benefit from the shift from oil-fired units to coal-fired units; on the contrary, they had to bear an extra burden to finance the rapid expansion of coal-fired units in the early 1980s.

In Table 3.3 we show the revenue requirements of CLP and HEC since 1979. We have attempted to work out the proportions of fuel cost, other operating costs (including depreciation and taxes) and permitted return in the total revenues of CLP and HEC. The proportion of fuel cost increased rapidly in the 1970s and the early 1980s due to the oil crises. After the introduction of coal-fired generators in the early 1980s, the proportions of fuel cost declined continuously. In 1994 CLP started to purchase electricity from its partially owned generating capacity in China, so the proportion of fuel cost in the total revenue fell significantly, from 21% to 13%. For HEC, the share of fuel cost in the total revenue in 1995 was 13%.

Table 3.3 also shows the proportions of fuel cost, other operating costs and the net return (i.e., profit after tax and transfers) for each unit (kWh) of electricity sold by the two companies in 1993. To control the effect of power purchased from China on cost structures, we use the year 1993 for our comparison between the two companies. In 1993, the introduction of coal-fired generators raised the profit margins of CLP and HEC to 32% and 43%, respectively. As fuel costs and other operating costs of the two electric utilities are similar, the higher price charged by HEC is mainly a result of its higher profit margin. If the lower price charged by CLP is due to its larger economies of scale, then the higher rate of return on capital enjoyed by HEC is simply a reflection of its smaller scale of operation.

In Table 3.4 and Figure 3.5, we show residential electricity

**Figure 3.5**
**Real Residential Electricity Prices in Asian Countries, 1973–1990**

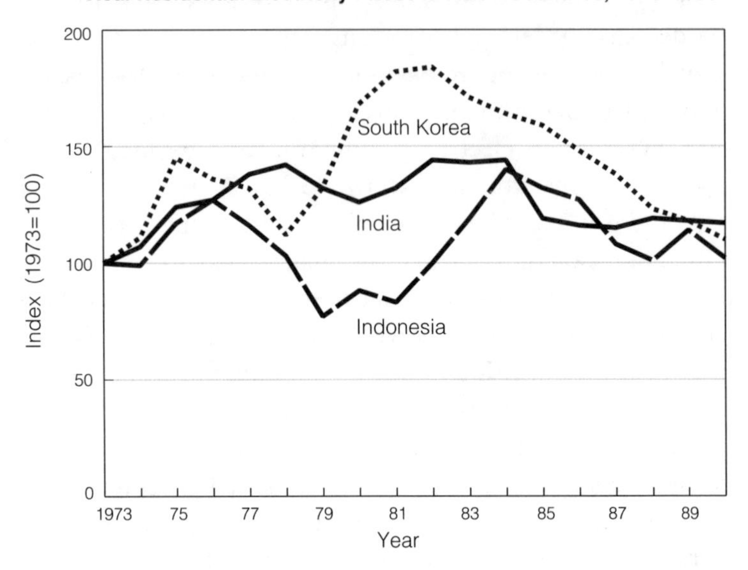


**Figure 3.6**
**Real Electricity Prices in Hong Kong, 1973–1990**

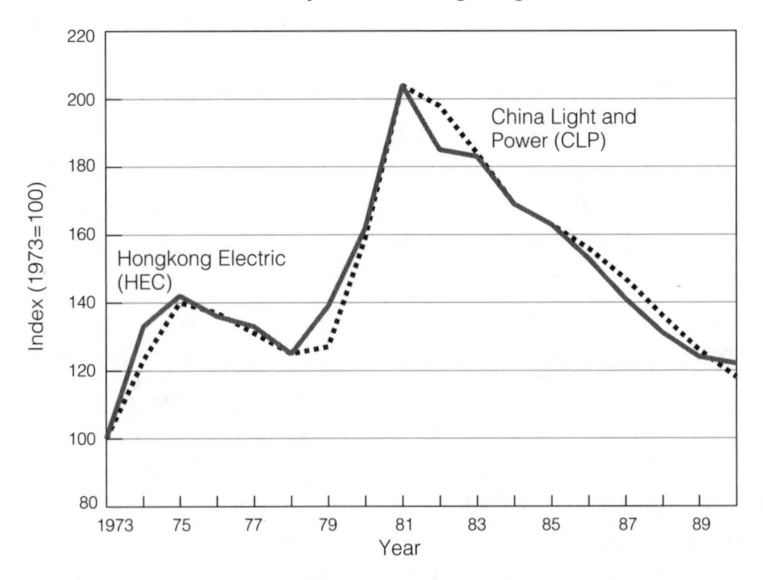

prices in several Asian countries for the period 1973–1990. As separate figures on residential prices are not available in Hong Kong, we have included the average electricity prices charged by CLP and HEC during the same period for the purpose of making our comparison (see Figure 3.6). In the four Asian countries selected for study, rising real prices of electricity in the early 1980s have been followed, to a greater or lesser extent, by price declines. The prices charged by utilities in Hong Kong show both the widest fluctuations and the greatest increase over the period.

As may be seen in Figure 3.6, electricity prices in Hong Kong fluctuated widely during the period 1973–1990. The construction of coal-fired generators in the early 1980s caused prices to increase sharply. But once these coal-fired generators had been commissioned, the two electricity companies were able to lower prices as fuel cost decreased. Despite this, the existing real tariffs charged by the two electricity companies remain at more or less the same levels of twenty years ago, right before the first oil crisis in 1973. In Table 3.5 we show the average tariffs and rates of return on equity in different Asian countries. Among the nine countries or cities, the average tariff in Hong Kong only compares favourably with the figures for the Philippines and Japan. Furthermore, there is also evidence that the price differentials between Hong Kong and other lower-tariff countries have widened since the early 1990s.

Finally, it should be noted that caution is advisable when making a direct comparison of the electricity prices in Hong Kong and other countries. Electricity companies in other countries operate under different economic and regulatory environments. Some utilities are privately owned, while others may be publicly owned. Population density and government taxes (or subsidies) also have a direct effect on electricity prices. Therefore, the relevant question should be whether tariff levels in Hong Kong could be lower. Although hydroelectric power is not available in Hong Kong, and the city is short of indigenous fuel supply, tariffs could be lower given Hong Kong's dense population and high level of urbanization.

### *On Returns and Capital Structure*

In Table 3.5 we find that the equity returns earned by electricity companies in Hong Kong are very attractive when compared with the utilities in other Asian countries. This positive performance in terms of return is partly due to the higher efficiency of the electricity companies in Hong Kong, but is also to a large extent the result of the regulatory arrangement in Hong Kong.

At present, the permitted rate on assets financed by equity is 15%, while the permitted rate on assets financed by the development fund is the same as that for debt-financed capital, both of which are fixed at 13.5%. Since the maximum interest rate payable on long-term loans is 8%, while the interest rate charged on the development fund is fixed at 8%, CLP and HEC tend to rely on debt-financing, as the permitted rate is in excess of the debt cost. This arbitrage process allows the shareholders of the two regulated utilities to earn a rate of return on their equity capital that is above 15%.

Under the rate-of-return regulation, the shareholders of a regulated firm will compare the permitted rate of return with the cost of equity capital to decide whether to invest or not. There are several approaches by which the cost of equity capital can be measured, the most notable one being the capital asset pricing model (CAPM). In the CAPM, the measure of market risk is known as beta (b). For example, the returns from an asset with a beta of 0.5 will change by 5% for each 10% change in the market's returns. It has been shown that the required risk premium for an asset is directly proportional to its beta. Therefore, the holder of an asset with a beta of 0.5 will require a risk premium only half as large as that offered by the market as a whole. If the market is efficient, in terms of competition, the cost of equity capital will be equal to the expected rate of return.

If a firm is financed in part by debt capital, then shareholders will bear financial risk as well as market risk. Finance theory argues that a higher debt-equity ratio will raise the equity beta of a firm, but will leave the firm's overall cost of capital unchanged. Finance theory also suggests that the overall cost of capital will not change if the firm changes its financial structure. Although the cost of debt capital is lower, the increase in debt raises the cost of equity capital, thus leaving the overall cost unchanged. The shareholders of a firm with a higher debt-equity ratio bear a higher degree of risk. Should the return of the firm decline, the rate of return on equity for a debt-financed company will decrease more sharply than will that for an all-equity firm, as debt holders have prior claims on the firm's income. In other words, a higher debt-equity ratio raises the variability of returns. The implication is that firms with higher debt-equity ratios will have higher beta values because of the increased financial risk.

But such an argument cannot be equally applied to electricity companies that are subject to the Scheme of Control. The existence of a development fund stabilizes each company's returns, and hence higher debt need not increase the risk of equity capital. Of course, there is a limit to the amount that can be drawn from the development fund. But so long as the development fund is able to compensate for any shortfalls in realized profits, the regulated firm is guaranteed earnings at the permitted rate of return. Fluctuations in the realized profit will not increase the risk borne by the shareholders. A direct testable implication thus follows: a higher debt-equity ratio will increase the rate of return on equity, but it will not raise the financial risk and beta of a regulated company.

Our analysis does not imply that the beta value of a company under the Scheme of Control will be zero if the size of the development fund is sufficiently large. This is because in reality the scheme regulates nominal rather than real returns. The company still has to face fluctuations in real returns resulting from unanticipated inflation. In the U.S., for example, several studies indicate that a rate-of-return regulation retards price adjustments during inflationary periods, resulting in a diminished financial performance and increased systematic risk for regulated firms (Norton 1985).

In Table 3.6 we show the debt-equity ratios and rolling betas (based on monthly data) of the two electricity companies for five-year periods. We can see that the debt-equity ratio of CLP increased from 0.71 to 1.43 and then decreased to 0.63, but its beta value remained relatively stable. There was a tendency for the beta value to decrease. However, while HEC's debt-equity ratio increased over time, its beta value decreased continuously. To conclude, empirical data have supported the notion that with the existence of a development fund, a higher debt-equity ratio does not imply higher financial risk. Instead, evidence shows that the higher debt-equity ratios of the two electricity companies were accompanied by higher rates of return on equity under the Scheme of Control.

## Government Policies on the Gas Industry

### The Consumer Council's Report on the Fuel Market

In Hong Kong's gas industry, however, there is no government regulation on returns and prices. The government believes that competition among alternative fuels drives prices down to their competitive levels. This long-held belief, however, recently received strong criticism from the Consumer Council of Hong Kong. In July 1995 the council published a report on the gas industry in Hong Kong, *Assessing Competition in the Domestic Water Heating and Cooking Fuel Market* (1995). The report (1995) attacked the government's safety policies and its lack of intervention, blaming them for helping the towngas monopolist, Hong Kong and China Gas (HKCG), to earn excessive profits, to the detriment of its customers. The report (1995) proposed a restructuring of the gas industry, with an aim towards introducing competition into the industry.

Despite the fact that HKCG has been earning "excessive profits" for years, the Hong Kong government refused to regulate the company, based on the belief that a choice of substitutes, such as electricity and bottled and piped LPG, are available to consumers. It has been argued by the government that competition among these alternative fuel suppliers lowers the prices charged by HKCG to their competitive levels.

However, after Governor Chris Patten took office in July 1992, this policy of non-intervention changed. In his Policy Address to the Legislative Council in October 1992, the governor called upon the Consumer Council of Hong Kong to work towards the development of a comprehensive competitive policy for Hong Kong. In response to the governor's request, the council set up a Working Group on Competition Policy to identify unfair, discriminatory, or anti-competitive business practices in Hong Kong, and to evaluate their effects on consumer interests. The industries identified by the council for review include supermarkets, banking, gas, property development, broadcasting, and telecommunications. After two years of study, the Consumer Council published its report on the study of the gas industry in July 1995. The report (1995) provided five major recommendations:

1. To promote competition in the gas industry, a common carrier system should be adopted. HKCG is obliged to open up its gas distribution network for use by other companies.
2. The government should encourage industry players to bring natural gas to Hong Kong, as it is cheaper, safer, and can be transported by the existing network.
3. Before a common carrier system becomes fully operational, HKCG should be subject to some government control. Instead of U.S. rate-of-return regulation or an arrangement similar to the Scheme of Control, a price-cap regulation is proposed.
4. To enhance the competitive power of electricity companies, property developers should be forced to offer three-phase electricity wiring together with gas piping, so that consumers can make a genuine choice between electric and gas for heating and cooking.
5. An energy commission to co-ordinate all energy issues should be established; the commission should be advised by an energy advisory committee.

## Sources of Market Power

To support its argument for government control and for the purpose of widening the market, the Consumer Council provides much evidence of HKCG's dominance within the current market, and of the abuse of its monopoly power. From the very beginning, the report (1995) argues, there has been imperfect competition among different energy suppliers in the market due to safety, technical, and cultural factors.

As was mentioned before, a turning point in the history of Hong Kong's gas industry, which was not discussed in the report (1995), occurred in 1981. After the publication of the *Report on the Safety and Legal Aspects of Both Town Gas and LPG Operations in Hong Kong* (1981), the market share of towngas increased dramatically (see Table 3.7). Apart from policies on safety which favoured the use of towngas over LPG, technical and cultural factors have enhanced HKCG's competitive power over electricity. A comparable alternative to the gas water heater is the instantaneous type of electric water heater. However, for technical reasons, such electric water heaters can only be installed where electricity supply is boosted with a three-phase electrical installation. As is argued by the Consumer Council, a three-phase installation is not required for common daily usage in most small to medium-sized apartments in Hong Kong. Hence, most electric water heaters installed in residential households to date are of a water storage type rather than an instantaneous type. The amount of heated water from the storage type is limited by its water storage capacity, and thus it compares unfavourably with gas water heaters, which can provide a continuous supply of heated water. The competitiveness of electricity companies is further reduced by a preference for "flame cooking" among the Chinese, who, as a result, favour gas for cooking. These technical and cultural factors, as argued by the Consumer Council, have caused imperfect or unfair competition between HKCG and electricity companies.

Although technical factor and cultural preferences favouring gas have given HKCG a competitive edge over alternative fuel suppliers, the same factors restricted the growth of the company in its early development. In the 1950s and 1960s, HKCG grew very slowly, compared with alternative fuel suppliers. Two factors accounted for this low rate of growth. Firstly, the supply of towngas was restricted to the areas where production plants and distribution networks were available. Alternative fuels like kerosene and LPG, however, could be delivered to nearly every customer by road transport. Secondly, towngas, which is "invisible", was not favoured by traditional Chinese families during this early period. They preferred other more "visible" sources of energy, such as charcoal and kerosene. Despite the unsuccessful introduction of towngas, HKCG continued to invest in its production and distribution facilities, and at the same time tried to promote its product to the community. When the living standards of Hong Kong people improved, the consumption of towngas increased remarkably.

From a historical point of view, the success of HKCG is to a large extent a result of the innovative attitude of the company's entrepreneurs. The high rate of return enjoyed by the company is partly due to Hong Kong's safety regulations and partly due to the reward, or risk premium, which has resulted from the company's innovations. In a market without entry restrictions, firms introducing unwanted products lose income and go out of business, whilst those that can successfully introduce new products are rewarded by returns higher than the competitive level. If competition really exists in the gas industry, there is no justification for the government to eliminate income arising from innovative activity, while not compensating losses suffered from unsuccessful innovations. Such asymmetric policies dampen the incentive to innovate.

Viewed from a different angle, however, if HKCG's success can be attributed to favourable policies (on safety) adopted by the government, and to the right to build transmission networks under public roadways without paying a commensurable price, there may be a case for government intervention. Government intervention is needed to prevent the company from using its monopoly power to exploit consumers. The benefits derived from its broad network should be extended to consumers and other fuel suppliers.

## The Government's Response to the Report

After studying the Consumer Council's Report for six months, in early 1996 the government decided not to impose any formal price or return control on HKCG. The government is of the opinion that even though the market share of HKCG has been increasing, there is no evidence that the company is abusing the power of its dominant position in the market. Prices charged by HKCG and its rates of return are in line with the two regulated electricity companies. Nevertheless, as the market share of HKCG is likely to increase in the future, the government agrees that there should be increased transparency in the company's tariff-setting mechanism through some formal consultative arrangements. In its response to the report, the government also proposes to commission a feasibility study on introducing a common carrier system in Hong Kong, and to form an energy advisory committee to advise on energy policy.

According to the assessment made by the government (Economic Services Branch 1996), HKCG is now capturing about half of the market for water heating and cooking fuel, and its overall market share is growing steadily, at the expense of LPG and electricity. If we only consider the gas market, HKCG has a market share of 66%, the rest belonging to LPG companies. Concerning prices in the fuel market, the average price charged by HKCG has been higher than that of piped LPG but less than that of electricity and cylinder LPG. The government has argued that as the difference between the prices charged by CLP and HKCG has narrowed in recent years, electricity companies have put competitive pressure on HKCG in the fuel market. Hence, some degree of competition for fuel exists within the local energy industry. In the next chapter, we shall compare the performance of these three utilities within the Hong Kong energy industry, and see to what extent government intervention (or non-intervention) has affected their operation and achievements.

## Industry Performance of Power Utilities

### Energy Prices

In Figure 3.7 we compare the tariffs charged by the three power utilities in Hong Kong. When comparing increases in electricity prices, the two electricity companies often consider the period since 1983, after the introduction of coal-fired generators. But as we can see, HKCG's price increases were not greater than were those of CLP and HEC, if we measure price increases over an extended period of time. Besides, towngas prices were lower than electricity prices when measured in terms of the same thermal unit. In real terms (based on 1973 prices), energy prices in Hong Kong in 1995 were at more or less the same levels as they were in 1973. In other words, real prices did not actually decrease over the period.

### Equity Returns

To determine whether HKCG is earning "excessive" profits, we can consider the equity returns earned by the company's shareholders and compare these equity returns with the cost of equity capital. The author has conducted a study on the costs of the capital of power utilities in Hong Kong, based on monthly data from 1977 to 1992. Data for the period after 1992 are not used, in order to avoid the accounting problems caused by the recent involvement of these power companies in property development and the asset revaluation of HKCG in 1993. The results are shown in Table 3.8. The risk factor (beta value) and the cost of equity capital (or required rate of return on equity capital) of HKCG is higher than CLP and HEC. This is not unexpected, as the returns of the two electric utilities are regulated by the Scheme of Control, and their development funds can stabilize the returns of the two companies. Based on our results, despite control imposed on CLP and HEC, all three power utilities earned returns in excess of their equity costs for the period 1979 to 1992 (Table 3.9). Therefore, government regulation may protect, rather than eliminate, the excessive returns earned by monopolists.

**Figure 3.7**
**Energy Prices in Hong Kong, 1973–1995**

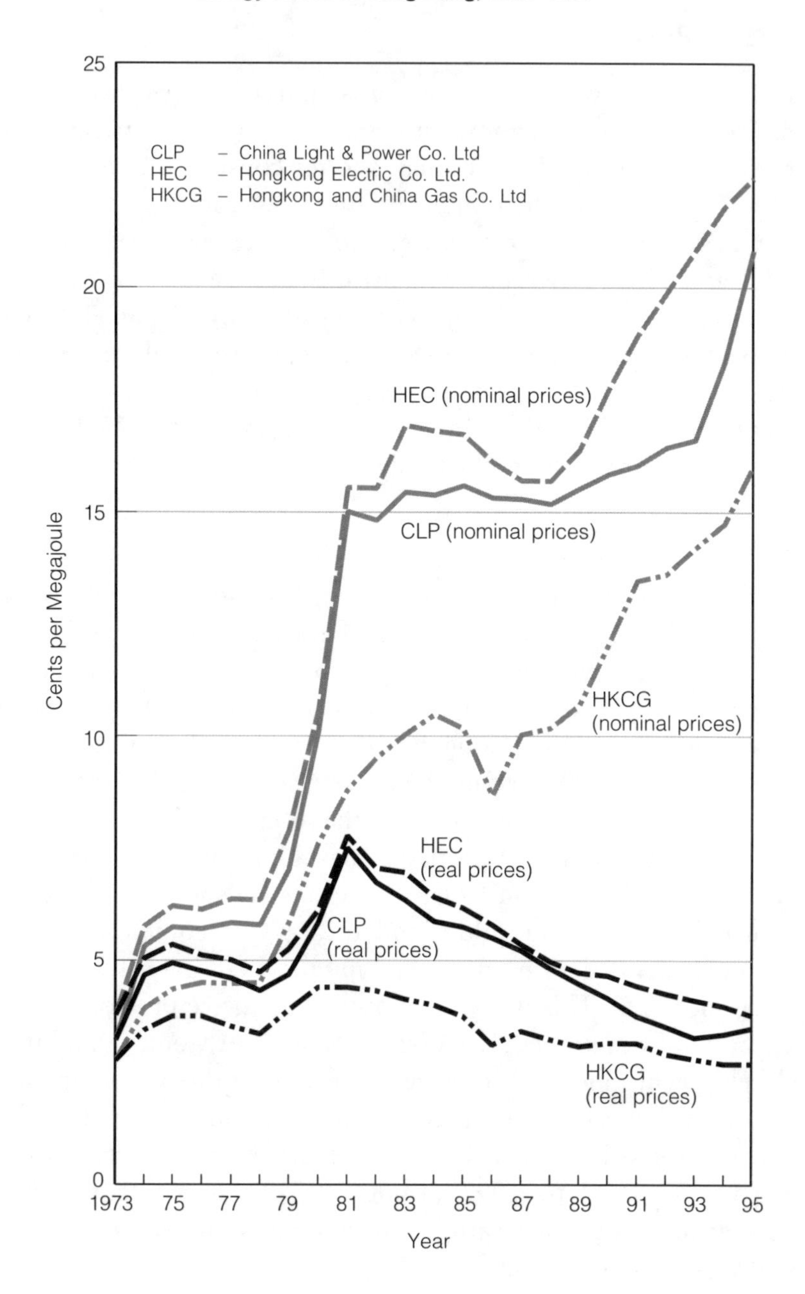

## Installed Capacity

Table 3.10 shows the expansion in installed capacity of the three power utilities since World War II. After signing the Scheme of Control Agreement with the government, CLP quickly installed new units in the old plant at Hok Un, and at the same time started constructing the new power station at Tsing Yi Island. On Hong Kong Island, apart from building new units at North Point, HEC started constructing a new power station on an outlying Island, Ap Lei Chau, in the 1960s. After the completion of the Ap Lei Chau Power Station, HEC decommissioned its plant at North Point and redeveloped it into a residential estate.

After receiving approval from the government to build a new power plant on Lamma Island, in the late 1980s HEC relocated its units from Ap Lei Chau to the new power plant. The vacant site was again redeveloped into a residential estate. In recent years, CLP has obtained approval from the government to redevelop its early power plant site at Hok Un. The government has also allowed the company to decommission its power station at Tsing Yi a few years ahead of schedule, and CLP is now negotiating with the government concerning the future use of the site. As mentioned earlier, HKCG is also planning to redevelop its gas plant site at Ma Tau Kok.

As a matter of fact, when the economy of Hong Kong expands and more new towns are developed, there will be a genuine need to remove the power plants from the city areas. However, decisions on relocation and early retirement of the power plants may mainly serve the interest of shareholders rather than of the consumers or the whole community. The shareholders may seek redevelopment when they consider that the property value is much higher than are the returns derived from electricity generation. Consumers and society might lose in the process if useful plants are decommissioned earlier than necessary and if some extra costs of removal and installing new transmission facilities after relocation have to be borne.

## Operating Efficiency

Figures 3.8 and 3.9 show the labour productivity of the power

**Figure 3.8**
**Customers per Worker of Power Companies**

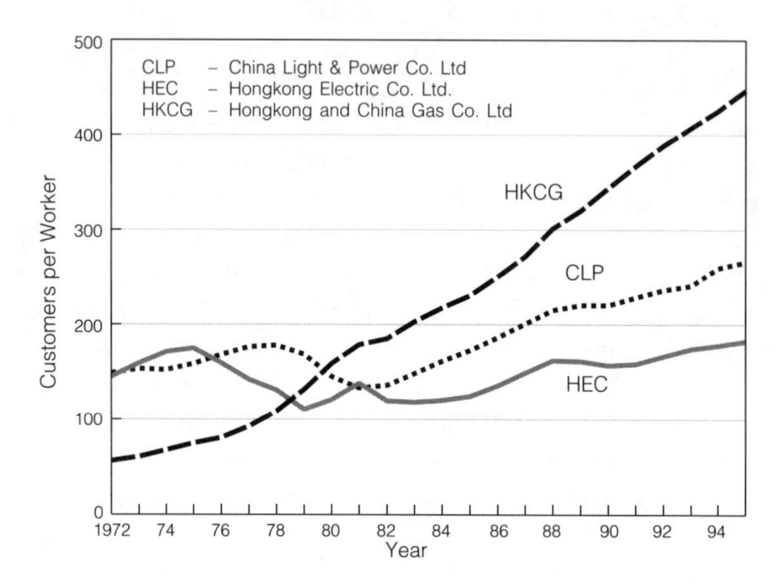


**Figure 3.9**
**Labour Productivity Growth (1973=100) of Power Companies**

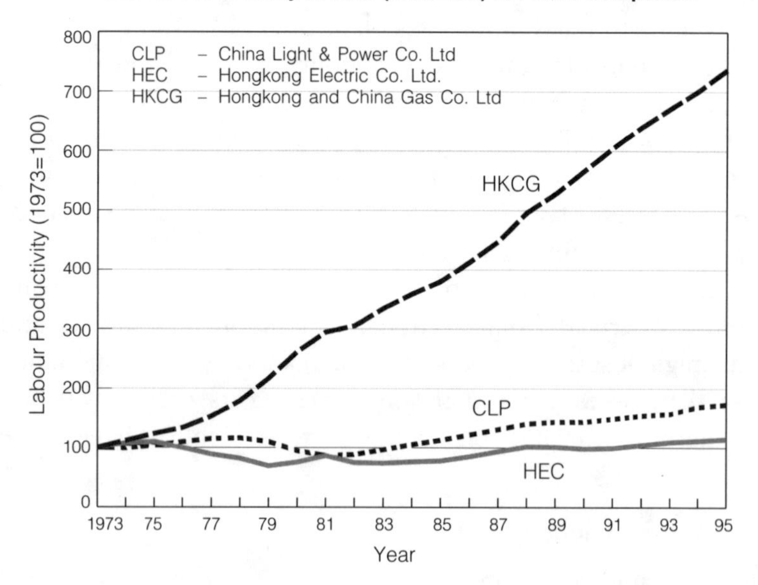

companies in Hong Kong. Despite improvement in technology and the introduction of more advanced production units, the labour productivity of CLP and HEC have been increasing very slowly. HKCG, the unregulated utility, outperforms the two regulated electric utilities in labour productivity growth. The labour productivity of the towngas company has shown remarkable improvement over time. But, as we find that real price of towngas in 1995 was only 1.9% lower than it was in 1973, it seems that the company has kept its productivity gain in higher returns and has not shared the benefit with its customers.

Table 3.11 shows the operating efficiency of the three power companies. The shift from oil-fired generators to coal-fired generators in the 1980s also increased the thermal efficiency of the two electricity companies to 36%. But, as was mentioned earlier, the use of coal-fired units has brought additional environmental problems, and extra capital investment has been required to reduce emissions. When CLP's new gas-fired units are commissioned in the coming years, its thermal efficiency will be further increased, as these gas-fired units have a thermal efficiency of 50%.

As we can see from Table 3.10, all three companies are now operating with high excess capacity. Reserve capacities of the two electricity companies are much higher than the international standard of 25% (see Figure 3.10). The excess capacity of HKCG is even greater, at 150% above its maximum demand. Energy users have paid higher prices to finance the over-expansion. In addition, the plant factor, load factor, and system loss of the two electricity companies have not shown any noticeable improvement over time (see Table 3.11). The plant factor and load factor of CLP deteriorated sharply in 1994 after it started purchasing electricity from China. The plant factor of HKCG also deteriorated as it expanded its production plant in Tai Po. Therefore, as a first step, better planning and co-ordination in the construction of new production units between power companies, and an improvement in load management, will certainly help reduce energy costs in Hong Kong.

Figures 3.11 and 3.12 show the monthly consumption of

Figure 3.10
Reserve Capacities of CLP and HEC

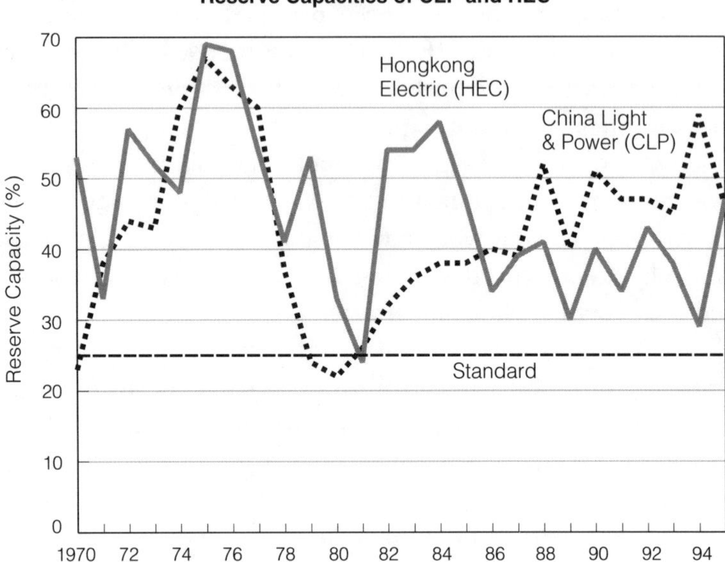


electricity and towngas in 1995. As can be seen from these two figures, the monthly consumption patterns of electricity and towngas are entirely different. Electricity demand has a summer peak, while towngas demand has a winter peak. In summer there is a greater electricity demand for air-conditioning. The increase in demand mainly comes from the commercial and residential sectors. The two electricity companies have to increase installed capacity to meet the summer peak. Demand for electricity falls in winter, but there is an increased demand for gas used for heating and cooking. If gas companies in Hong Kong can successfully promote their businesses and lower the demand for electricity during summer, this will help reduce the pressure to install new electricity-generating facilities.

## Capital Expenditure

Table 3.12 shows the capital expenditure of the three power companies. According to the financial arrangement between CLP

**Figure 3.11**
**Monthly Consumption of Electricity in 1995 (thousand)**

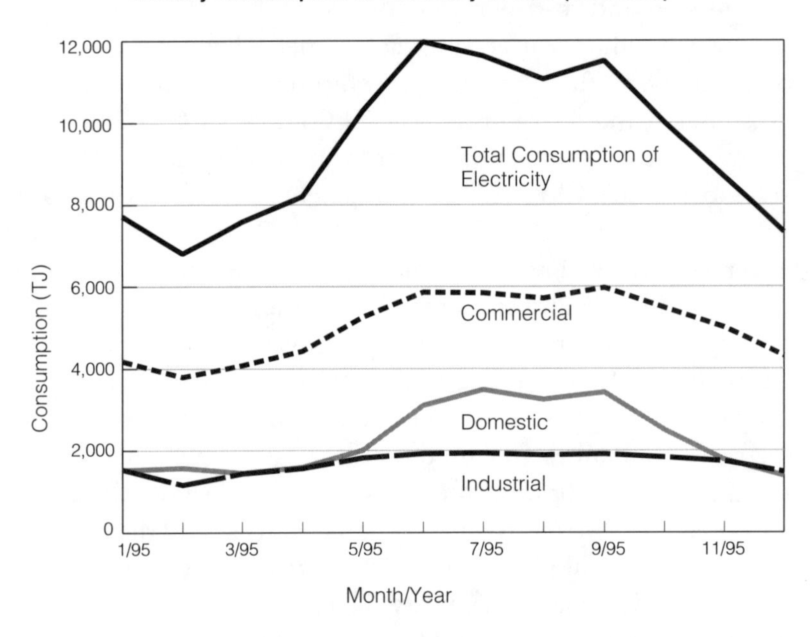


**Figure 3.12**
**Monthly Consumption of Towngas in 1995**

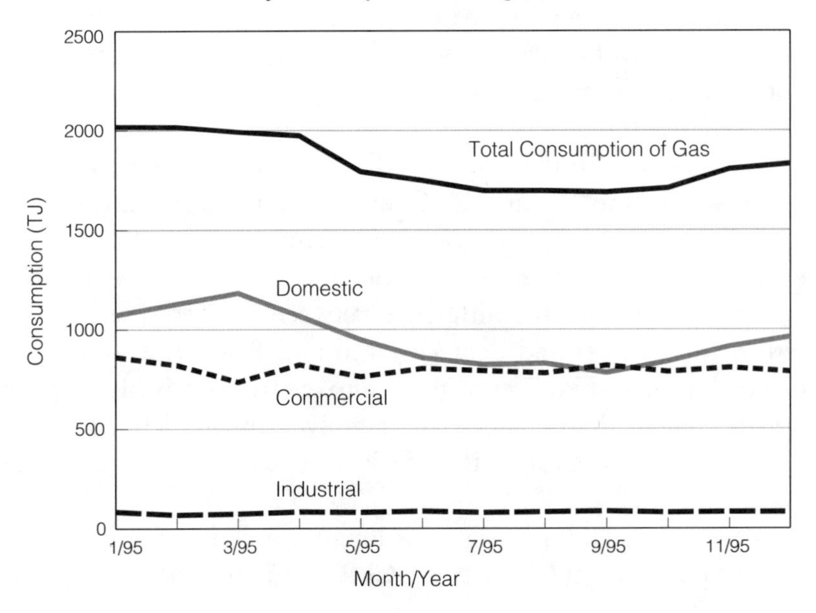

and its partner, Exxon, CLP is responsible for constructing a new transmission and distribution network, and the two partners co-operate in building new generating facilities. Therefore, we are able to distinguish the capital expenditure on generation from transmission and distribution. But HEC and HKCG do not supply separate data on production and transmission. Of the total capital cost incurred by CLP and Exxon, generation cost takes a larger share, as compared with transmission cost. Over the years, all three power companies have built more and more expensive (in real terms) production and transmission facilities.

Although better facilities can improve supply reliability and reduce fuel cost and energy loss, energy users may suffer if a company over-invests. As the permitted returns of the two electricity companies are based on fixed assets, customers have to pay higher prices for unwanted excess capacity. Under the Scheme of Control, the two electricity companies can expand their capacity once their financial plans have been approved by the government, regardless of whether the capacity is later found to be useful or not. Hence, the Scheme of Control has put the burden on the regulator to approve or disallow expansion plans submitted by the regulated utilities. The towngas company has also charged higher prices on captive users to finance its over-expansion. Since the company is not under government regulation, its expansion plan is not subject to public scrutiny.

Furthermore, as the returns of the two electricity companies are protected by the Scheme of Control, financial institutions are willing to lend them money at attractive rates to finance the expansion. If the two companies can successfully issue debt, they will be able to earn the difference between the permitted rate of return (of 13.5%) and debt cost (around 8%). The Scheme of Control seems to be a licence granted to the two companies to print money. Their shareholders can simply rely on debt (and the development fund), rather than their own equity capital to finance expansion.

At present, CLP has signed a long-term contract to purchase nuclear power from China. Since CLP's return on local generating facilities is protected by the Scheme of Control, the company will maximize its power purchase from China. The company will then derive income from its investment in the nuclear power station, which is not subject to the Scheme of Control and can continue to earn the same permitted return from its assets protected by the scheme. Since CLP will take electricity from China first, its local generating facilities will be running with high reserve capacity should the growth in local demand for electricity slow down. The existing regulatory arrangements seem to protect the company's investment in China and Hong Kong.

Some mechanisms to prevent over-expansion of production facilities and to remove protection on unwanted capacity must therefore be put into place. Energy users should be allowed the choice to enjoy cheaper energy available from other suppliers. Hong Kong can borrow the experiences of market liberalization from other countries to improve the efficiency of the energy industry.

# CHAPTER 4

# Experiences of Liberalization in the Energy Industry

## Competition in the Energy Industry

In recent years, liberalization in the utility industries has been a major theme of public policy in many countries. This liberalization trend, that is, the introduction of competition, has often come with a restructuring of the industry to remove barriers to entry. It has been shown that the forces of competition can better protect consumer interests than can extensive regulation of monopoly. But this does not mean competition and monopoly regulation are perfect substitutes. In some situations, government regulation is still needed to guard against anti-competitive behaviour by the dominant firms in a newly liberalized market. In this chapter, we review the experiences of market liberalization in the energy industries of the U.S. and the U.K. We consider the attempts made by the regulators in these countries to have the barriers to competition in the energy industry removed.

## The U.S. Energy Industry

### The Electric Power Industry

At present, electric utilities in the U.S. can either be privately or publicly owned. In its early period of development, the U.S. electric power industry consisted of a large number of small producers.

Each producer generated electricity, which was distributed to a relatively small surrounding area (Asch and Seneca 1989). By the 1900s some of these producers had achieved local monopoly status by obtaining municipal franchises (Priest 1993), while others, taken over by municipalities, became publicly owned firms. Individual state commissions subsequently regulated the privately owned firms.

Until 1910 the number of electric power systems in the U.S. increased steadily. With technological developments, companies were merged to exploit scale economies in the generation and transmission of electricity. This led to a decrease in the number of electric systems. The decrease was also caused by the merger movement for financial gains and the growth of holding companies. In the 1920s the pyramiding nature of holding companies enabled the owners of the company at the top to control a large amount of assets with a relatively small initial investment. This greatly increased the risk of investment, as most of the capital of the holding companies was raised by debt financing. As the holding company was separate from the operating companies, it could relieve its owners of any financial responsibility caused by the latter's losses. The collapse of the stock market in 1929 converted many of the holding companies' inflated profits into huge losses. To prevent the recurrence of such financial abuse, the Public Utility Holding Company Act 1935 (PUHCA) was enacted. The Federal Power Act (FPA) was passed in the same year. Tighter restrictions on utility mergers were then imposed. The PUHCA prohibits or limits investment by utility holding companies into non-utility businesses. It also prohibits the acquisition of distant utility companies (Hempling 1995).

Because most utilities were too small to exploit all available economies, they engaged in some co-operative activities such as joint ownership of facilities, interconnection, and power pooling, in order to obtain certain advantages. Power pooling refers to formal and informal agreements among independent utilities to co-ordinate some, or all, of their investment and operating activities.

Control over interstate wholesale sales of electric power made by privately owned utilities rests on the Federal Energy Regulatory Commission (FERC). Formal pooling agreements must also be approved by the FERC.

During the 1970s, despite rapidly rising electricity prices, the real earnings of electric utilities declined, their stock prices fell, and their bond ratings were reduced. Regulatory commissions were criticized for failing to offer adequate rates of return to compensate investors for the capital they had to provide. There was also a fear of insufficient electricity supply as electric utilities reduced their investments (Joskow and Schmalensee 1983). In addition, the debates over deregulation had become intense by the late 1970s. The Organisation of Petroleum Exporting Countries (OPEC)-induced "energy crisis" of the mid and late 1970s also intensified. The congress then passed the Public Utility Regulatory Policies Act (PURPA) in 1978 in its attempt to escape the crisis (Stalon 1995). The PUPRA required, among other things, utilities to buy power at regulatory-determined rates from co-generators, small power producers, and renewable fuel users whose generating facilities were found by the FERC to be "qualified facilities" under requirements of the law. The PUPRA thus provided an institutional basis for encouraging energy conservation and competition and increasing energy supply through the co-generation of electricity.

Co-generation is the process of generating electricity as a by-product in the manufacturing of other goods, such as paper or steel. By-product electricity can be used by its producers and/or sold to another party (Fox-Penner 1990). The PUPRA guarantees "qualified facilities" the right to sell self-made power back to the nearest utility at the latter's full "avoided cost". FERC defined "avoided cost" as the incremental cost to an electric utility of electric energy or capacity, or both, which, but for the purchase from the co-generator, would generate itself or purchase from another source (Berry 1989). Although co-generation technology predated PURPA by many decades, the act broke the monopoly bottleneck in the transmission and encouraged competition in the generation of electricity. To enhance competition, co-generating facilities cannot be more than 49% utility owned. Because the avoided cost of monopoly utilities was so much higher than was the cost of electricity from co-generation, the PURPA policy led to a massive oversupply in generating capacity and stabilized prices of electricity in many states during the 1980s (Summerton and Bradshaw 1991). The PURPA also directly and indirectly encouraged self-generation and wheeling. Since all states except Texas have an interconnected grid, FERC retains regulatory jurisdiction over wholesale wheeling.

To encourage further competition in the generation business, the Energy Policy Act of 1992 (EPAct) lifted some of the restrictions imposed by the PUHCA of 1935 on utility diversification and created a new category of corporation, "exempt wholesale generators" (EWGs). EWGs are defined roughly as companies that own generators and sell electricity exclusively at the wholesale level anywhere in the world or at retail level outside the U.S. As these new power producers are not considered "electric utilities", they are exempt from most PUHCA requirements. The EPAct also gave FERC broad authority to mandate wholesale transmission access. Combined with the actions of state commissions to restructure the electric industry, the act enhances competition at the wholesale and retail levels. The act made it possible for retail customers to exit from their franchised utility and buy electricity from other areas at more attractive prices.

The restructuring process in the U.S. electric power industry resulted in the problem of stranded costs (investment rendered useless because of open competition) incurred by franchised monopolists. There is still a debate about whether these monopoly utilities should be compensated for their loss of revenue. If compensation is required, there is also a question of the extent to which they are to be compensated. It is proposed that an efficient exit fee be charged to those customers who exit from their franchised utility. As the existing transmission systems are owned by different entities, the opening up of the transmission grid for equal access has also given rise to issues of ownership and control of transmission.

## The Gas Industry

The deregulation of the natural gas industry in the U.S. dates back to the 1970s. Faced with an energy crisis induced by price control on gas at the wellhead, congress passed the National Energy Act of 1978, which included the Fuel Use Act, the Natural Gas Policy Act (NGPA) and the Public Utility Regulatory Policy Act (PURPA), mentioned above. To provide producers with enough incentives to explore and develop new gas fields, the NGPA deregulated the price of newly discovered gas.

This policy of partial price deregulation, however, led to the rapid rise in retail gas prices during the early 1980s. Because new supply costs were averaged against the old gas costs retained at pre-NGPA price levels, the average price of gas increased as the more expensive, newly discovered gas constituted an increasingly large percentage of the pipelines' average cost. It was argued that price levels were well above those that would exist in a competitive market. As a result, consumption of gas decreased when large customers switched to alternative fuels and a conservation policy took effect. To mitigate the emerging problem of gas surplus, gas producers provided take-or-pay contracts at discounted gas prices to pipelines in return for increased sales. But such an arrangement was considered by courts as discriminatory, and this caused a shift in the function of pipelines from that of merchants of gas to that of transporters of gas (ANR 1994).

In 1985 the Federal Energy Regulatory Commission (FERC) issued Order 436, which required pipelines to provide equal access to anyone requesting transportation of gas on a non-discriminatory basis. The restructuring of the pipeline business from a merchant function to primarily a transportation function culminated in 1992, when FERC issued Order 636. The stated objectives of Order 636 were to increase competition and to allocate pipeline capacity more efficiently. Total price deregulation of gas supplies at the wellhead was also completed by January 1993.

Order 636 aimed to complete the transition to a competitive natural gas industry instigated by NGPA in 1978. It was believed that Order 636 would allow all natural gas suppliers, including pipeline companies acting as sellers of gas, to compete for gas purchasers on an equal footing. In the preamble of Order 636, FERC concluded that bundled service was the cause of a lack of comparability between an open-access transportation service and the bundled service provided by pipelines. Pipeline companies were accused of using their power of providing bundled service (gas sales and transportation) to enjoy a competitive advantage over other market participants. From November 1993, pipeline companies had to render transportation service on an equal basis to all shippers of gas. Natural gas could then be purchased from a variety of third-party suppliers and transported by pipeline companies under some firm transportation arrangements. Suppliers of gas in the market were to provide a list of separate services, including metering, storage, transportation, maintenance and repairs, etc., from which their customers could choose. The unbundling of gas services, though facilitating market competition, inevitably raised the transaction costs in the market. Order 636 also provided a competitive environment for pipeline companies to use a bidding process to assign the excess firm transportation capacity.

As a result of Order 636, a pipeline's ability to balance gas demand and minimize costs has been greatly reduced, as it no longer manages all the capacity and storage on its system as an integrated whole. Order 636 permits pipeline companies to recover their costs on "stranded investment" (facilities rendered unnecessary by Order 636) from their customers (Gorak and Ray 1995). The majority of these costs are paid by local distribution companies, which passes them to their captive customers. The new industry structure in the gas industry has enhanced the role and responsibility of regulators. They are now charged with ensuring that each state will receive the benefits from a policy of open-access transportation and reliance on market forces in the gas industry.

## Regulatory Procedures in the U.S.

### Rate-of-Return Regulation

In the U.S., state public utility commissions regulate the price and non-price terms and conditions of retail electricity sales, while the Federal Energy Regulatory Commission regulates wholesale sales made by privately owned utilities. In most cases, privately owned or investor-owned utilities operate as franchised monopolists serving retail customers in legally defined areas. The franchising process and the terms of franchises vary from state to state. In a few areas, investor-owned utilities have overlapping franchises and, at least in theory, can compete with one another for customers (Joskow and Schmalensee 1983). Most of these privately owned utilities are subject to a rate-of-return regulation.

Much like those of the Scheme of Control, the objectives of the rate-of-return regulation are to protect the consumer and at the same time provide the company with a "fair" rate of return (Crew and Kleindorfer 1987). The regulatory process begins with a formal request made by a public utility for a change in the level or structure of its existing rates, accompanied by a submission of evidence in support of the request. The utility must file its case and be prepared to be examined on it. Following the filing, the regulatory commission presides over a formal proceeding in which the evidence provided by the utility, along with evidence submitted by other parties such as commission staff and customers, is presented and examined. Intervenors, principally the state-appointed Rate Counsel, have the opportunity to object to the case, and then file their own testimony criticizing it. The two sides, public utility and intervenors, may either discuss their differences and agree to a "stipulation" which has to be approved by the commission, or to litigate the case before an administrative judge of the law.

The litigation is similar to normal court proceedings in which cross examination, rebuttal, and the like take place. After the hearings, the judge prepares a report, which will go to the commission for a final decision. In some cases, the commission may, on its own initiative, order an investor-owned utility to change the level and structure of its rates if the rates are found to be inconsistent with state law.

## Problems of Rate-of-Return Regulation

Crew and Kleindorfer (1987) identify four categories for evaluating economic efficiency: allocative efficiency, productive efficiency, dynamic efficiency, and governance cost efficiency.

Under rate-of-return regulation, prices are based on average cost rather than on marginal cost. Hence, except with the use of an optimal two-part tariff, the rate-of-return regulation can only achieve optimal second-best prices, at most. In addition, such a regulation is similar to a cost-plus contract, which provides little incentive for regulated firms to minimize costs and achieve productive efficiency. This is because when they successfully reduce costs resulting in their returns exceeding the allowed level, prices will be reduced in the next review. The rate-of-return regulation has been criticized for encouraging utilities to over-expand; this is the so-called Averch-Johnson effect (Averch and Johnson 1962). If the allowed rate-of-return on capital is above the cost of the capital, then the regulated firm will substitute capital for other factors of production and operate at an output where cost is not minimized.

Apart from its deficiency in productive efficiency, the rate-of-return regulation is also criticized for its poor dynamic efficiency. A system is dynamically efficient if it is able to encourage the regulated firms to adopt innovation and invention, and to accommodate changes in tastes and preferences. Under the rate-of-return regulation, profits are fixed by the regulators. Regulated firms are not rewarded for taking the risks of innovation; they have few incentives to adopt cost-saving innovation or to introduce new products. Governance costs are also substantial under the rate-of-return regulation, as it involves considerable resources for the process of administrative hearings and litigation. With the huge amount of monopoly returns and administrative expenditure at stake, the rate-of-return regulation also encourages wasteful rent-seeking activities taken by both the monopolists and regulators (Crew et al. 1987).

## Incentive Features of the Rate-of-Return Regulation

Joskow (1986) argued that theoretical and empirical studies on the Averch-Johnson effect failed to consider the actual regulatory process. In practice, regulatory commissions are not bound to set rates that cover all costs incurred by regulated firms. Regulators have the authority to "disallow" capital costs (and any other costs) to be included in setting rates, if they find that the associated expenditures are imprudent or unnecessary. This "threat of disallowance" limits expenditure on "gold plating" (unnecessary capital expenditure), and provides at least some incentive for the utility to make efficient investment decisions. However, the "threat of disallowance" or "prudence review" may cause under-investment (Gilbert and Newbery 1994) and discourage the use of innovative technology (Lyon 1995).

In practice, the rate-of-return regulation does not take place in a continuous fashion. Prices are set for an interval of time, during which the utility is free to choose whatever input combination it likes, until the next regulatory review. Bailey and Coleman (1971) found that the existence of regulatory lag induced utilities to reduce a capital bias. Baumol and Klevorick (1970) also found that regulatory lag allowed utilities to keep the benefits of improved cost efficiency until they were asked by commissions to reduce prices in the next review. Joskow (1974) found that there was an asymmetry in the formal regulatory review. Utilities tended to ask for price increases when costs rose, but when costs fell, they were often allowed to keep the existing prices and enjoy higher rates of return. Hence, as a result of the regulatory lag, the actual rates of return earned by utilities could be above or below the commission-determined fair rate of return.

## Rate-of-Return Regulation and the Scheme of Control

Do the above incentive features of rate-of-return regulation exist in Hong Kong? Although electric utilities in the U.S. and Hong Kong are both under the rate-of-return regulation, there are some differences in the regulatory procedures. First, utilities in Hong

Kong are under formal long-term contracts, and the nominal permitted rates of return are fixed for fifteen years. In the U.S., the nominally allowed rate of return is not fixed. When utilities think that the existing prices do not allow them to earn a fair rate of return on investment, they have to file formal requests for price increases. The nominally allowed rate of return depends on the results from financial models and other evidence (such as inflation rate) provided by a company to support its request. Second, there are no independent regulatory commissions and public hearings in Hong Kong. The regulatory duties are shared by various government departments. Third, so long as a balance exists in the development fund, the electric utilities in Hong Kong are guaranteed the permitted returns every year. In Hong Kong, regulatory lags do not exist; any return in excess of the permitted return will be transferred to the development fund automatically.

An important distinction between the two rate-of-return systems is on the determination of the allowed rate of return. In the U.S., the allowed rate is based on a weighted average cost of capital, which takes into account the cost of equity and the cost of debt. The weights depend on the debt-equity ratio. In Hong Kong, the government did not explain how the permitted rates of return were set (at 13.5% on debt and development fund capital and 15% on equity capital) and why the duration of the contract was fixed at fifteen years. However, as we have considered, once the permitted rate of return based on fixed assets is set, the electric utilities can earn returns on equity in excess of the permitted level by adjusting the modes of financing. In other words, the weighted average rate of return on capital assets is fixed, but the rate of return on equity earned by the regulated utilities is not.

When criticizing high governance costs of the system in the U.S., we have to be aware of the trade-off between governance cost efficiency and transparency. When compared with the existing system in Hong Kong, the U.S. system is more transparent, thus giving customers better access to information and enabling them to argue against price increases. In Hong Kong, there is lower degree of transparency and there are no public hearings on rate cases. Price increases by regulated monopolists are granted more or less automatically. This regulatory process, though less adversarial in public, can be costly if it arouses protests by consumer groups and grass root organizations (Liu 1991).

As the major concern of the rate-of-return regulation is one of equity and fairness, it is not surprising that it produces some inefficient outcomes. The many shortcomings of the rate-of-return regulation prompted interest in incentive regulation in the 1980s. In recent years, theories of regulation have been generated mainly in the area of incentive contracts between regulated firms and the government. In the next section we will consider one of these incentive contracts, namely the price-cap regulation, which is now widely used in U.K. privatized industries.

## U. K. Price-Cap Regulation

### Price-Cap Regulation in the Electricity Supply Industry (ESI)

After the Conservative Party came to power in 1979, there was a shift in the U.K. energy policy. The 1983 Energy Act attempted to introduce some competition into generation and supply by liberalizing third party access to the then nationalized transmission and distribution networks. The Area Boards, which were public corporations responsible for the distribution of electricity in their regions, were also required to buy electricity from sources other than the government's Central Electricity Generating Board (CEGB). But the act had little effect on the ESI: no private generator made any form of arrangement to sell electricity to a distant consumer through the networks. The act failed to attract new entrants because of the anti-competitive behaviour of the CEGB in adjusting the cost or price (Vickers and Yarrow 1988, 1991a). In February 1988 the British government published two White Papers to restructure the ESI in England and Wales. The main features of the new structure were:

1. The division of non-nuclear power stations owned by the CEGB between two new companies: National Power and PowerGen.
2. The sale of the twelve Area Boards, which were renamed Regional Electricity Companies (RECs) and became licensees in March 1990. RECs have two separate businesses: distributing electricity from the grid and supplying electricity to consumers. Entry to supply is free, while entry to distribution is not.
3. Transmission is separated from generation, and is carried out by the new National Grid Company (NGC), which is jointly owned by the twelve privatized RECs (it became a separate company later). The wires owned by NGC and the RECs must be made available for use by third parties at regulated prices. The NGC is responsible for despatching power stations in accordance with arrangements for the power "pool" (under the Pooling and Settlements Agreement), which is the new wholesale market for power. RECs can enter into generation to a limit of 15% of their requirements. The entry to generation is free.

The framework for regulation later appeared in the 1989 Electricity Act. The act set up the Office of Electricity Regulation (OFFER) to regulate the industry. Its head is the Director General of Electricity Supply (DGES). Professor Stephen Littlechild, who first advocated price-cap regulation, is OFFER's first director. His main duties are to ensure that demands for electricity are met and that licensees can finance their activities, to monitor licence conditions, and to promote competition in generation and supply. Licence conditions can be changed either by agreement with the licensee, or by the DGES making a successful reference to the Monopolies and Mergers Commission (MMC), the U.K. competition authority (Vickers and Yarrow 1991a). The price-caps imposed on the privatized firms are as follows:

1. The price charged by the National Grid Company for the use of its system is capped according to an "RPI minus X"

formula. The X factor in the cap is determined in such a way as to allow a regulated firm to share the benefit from increased efficiency. If the productivity of the regulated firm is increased at a rate faster than the X factor, then the firm can keep the excess returns. The excess returns will only be eliminated in the next regulatory review.

2. The price charged by each REC takes the form of RPI plus X, with the X terms varying among the RECs (depending on their investment requirements).
3. The price charged to consumers with a peak demand of less than 10 MW is subject to RPI – X + Y regulation. The Y factor reflects the cost to the REC of electricity purchases, transmission and distribution, and a fossil fuel levy. These cost components are beyond the control of the REC.

Figure 4.1 shows the structure of the privatized electricity industry in England and Wales. As we can see, the British government has adopted a structure of vertical separation by splitting the generation, transmission, and distribution of electricity in England and Wales into separate businesses. This new structure aims to promote competition in the generation and supply of electricity and to prevent abuse of market power in transmission and distribution by means of a price-cap regulation. Competition in the supply business, which was initially restricted to large users, was intended to be expanded gradually to small users.

**Figure 4.1**
**The Privatized Electricity Supply Industry in England and Wales**

| CONTROL | STAGES | | | |
|---|---|---|---|---|
| | Generation | Transmission | Distribution | Supply |
| Regulation | nil | RPI – X | RPI + X | RPI – X + Y |
| Entry | Free | nil | nil | Increasing |

Under the new industry structure, most electricity is provided on long-term hedging contracts between generators and RECs. As demand for electricity is so volatile, the price of marginal supply is set by the pool auction, operated by the NGC. The operator of the pool forecasts demand a day ahead on a half-hourly basis. It lists the bids from generators for providing electricity to the pool, and then agrees to pay for the capacity required. Theoretically, to meet the demand, the NGC should accept the cheapest bids first, and then proceed down the list to the most expensive.

However, it is evident that the duopolists (National Power and PowerGen) in power generation have exercised their market power to raise pool prices above marginal costs (Armstrong et al. 1994). They have little incentive to undercut each other. To reduce their power in setting pool prices, the Director General of OFFER has forced National Power and PowerGen to sell 10% of their capacity to independent generators. In addition to the problem of pool pricing, competition in generation brings concern about excessive investment in the building of gas-fired turbines. Following the privatization of the RECs, National Power, PowerGen and the NGC in England and Wales, the two vertically integrated electricity companies in Scotland, Scottish Power and Scottish Hydroelectric, were also privatized in 1991.

Preliminary conclusions drawn over the past few years have also suggested that competition in the supply market has caused a re-balancing of the tariff structures in favour of larger industrial users (Yarrow 1994). In the first three years after privatization, the price of electricity actually rose in real terms for most consumers, while the operating profits and rates of return of the two generators increased rapidly (*Economist* April 13, 1996).

## Price-Cap Regulation in the Gas Industry

In December 1986 British Gas (BG) became the second public utility company in the U.K. to be privatized after British Telecommunication (BT). In order to speed up the process of privatization, the British government did not restructure BG's business. BG remained a virtual monopolist in the transportation and supply of gas. It was also the only U.K. purchaser of gas from gas producers. The privatization was accompanied by the creation of a new regulatory body, the Office of Gas Supply (OFGAS). Price-cap regulation was imposed on BG in the small customer market, while there was no explicit regulation in the large customer market. The government believed that competition from other fuels and the policy of open access of BG's pipeline network would be adequate to reduce gas prices to their competitive levels (Vickers and Yarrow 1988). The British government was alone in believing that competition would be successful with a monopolistic structure. Virtually all other commentators, including the government's own supporters, disagreed with this.

Experience after privatization has shown that the industry structure did not foster the creation of a level playing field for other competitors. BG was accused of practising extensive price discrimination. BG's contracts with non-tariff customers, which were almost all industrial and commercial customers, were confidential. Prices were individually negotiated, and they were set higher for those customers less well-placed to use alternative fuels (Armstrong et al. 1994). BG also used its dominant position in the supply market to act strategically in its gas-purchasing policy, aimed at foreclosing entry.

After identifying the anti-competitive behaviour of BG, the Monopolies and Mergers Commission (MMC) changed its policy from relying on competition from other fuels to promote gas-to-gas competition. The MMC report of 1988 required BG to publish its price schedules and prohibited price negotiation and discrimination. The report also requested that BG publish more information about access terms on its pipelines, which would provide more details concerning transmission and distribution costs to potential competitors in the supply market. When determining the transportation prices, a fair rate of return on BG's pipelines was decided by OFGAS.

In 1991 the Office of Fair Trading (OFT) conducted a review on competition in the gas industry. The review found that most of the gas contracted for by BG's competitors was used for electricity generation. Competition in the non-tariff gas market was still limited. OFT concluded that the obstacles to competition were BG's monopolistic position in the tariff market and its control of transportation and storage facilities. These resulted in cross-subsidization and price discrimination against competitive suppliers. OFT proposed full divestiture of BG and reduction of its market share in the non-tariff market, but was willing to accept the creation of a separate transportation subsidiary as a compromise (Armstrong, et al. 1994). In response to the OFT review, BG agreed that OFGAS would regulate its access charges but failed to reach agreements on the allowed rate of return.

Faced with a tougher price-cap in the tariff market and a forced reduction in its market share in the non-tariff market, BG referred its business to the MMC for arbitration. MMC published its report in 1993. Among its recommendations, BG was again requested to separate its trading business from transportation. But the government did not accept the divestiture proposal. BG was allowed to retain ownership of trading and transportation, but had to operate them as separate subsidiaries with separate accounts. The government also decided to bring forward the ending of BG's monopoly over the tariff market in the south-western part of England in 1996. A choice of gas supplier for all consumers over the whole country is planned for 1998 (the same year in which a similar choice of electricity suppliers is planned to go into effect).

In February 1996 BG announced a radical demerger. The company decided to separate its energy exploration business from its gas supply. TransCo International is to handle international business, pipeline work, and exploration activity, and British Gas Energy (BGE), which engages in some offshore gas production, is mainly to serve the company's 19 million domestic customers. The Department of Trade and Industry approved both companies to be listed on the London Stock Exchange. The demerger, which would come into effect in 1997, was claimed to aim at preparing BG for greater competition in the British gas supply market.

## Arguments for and against Price-Cap Regulation

In their 1989 paper, Beesley and Littlechild summarized the arguments for and against a price-cap regulation

First, they suggested that RPI–X is more efficient than a rate-of-return regulation. It is less vulnerable to "cost-plus" inefficiency and over-capitalization (the Averch-Johnson effect). This is because the regulated firm can keep whatever profits it earns during the specified period, but it must also accept any losses, thus there is an incentive to produce as efficiently as possible. The intention of price-cap regulation is to mimic a Schumpeterian-type competitive process in which more efficient firms can keep excess profits for a period until the next price review. As efficiency increases, it is possible for both the company and the consumers to benefit. The company enjoys higher profits, while the consumers buy at lower prices. Since regulated companies keep all profits, they have a greater incentive to innovate and to introduce new products, resulting in a higher dynamic efficiency than that realized under the rate-of-return regulation.

Second, RPI–X allows the company greater flexibility in adjusting the structure of prices, as there is no price control on services outside the basket. As a result, the regulated firm can adopt pricing arrangements to achieve optimal second-best pricing. Within the specified period under the existing price-cap, a firm can learn the cost structure and then adjust the price structure accordingly.

Third, it is argued that RPI–X is easier to operate and that its governance costs are lower than the U.S. rate-of-return regulation. Both regulators and company need to devote fewer resources to operating the system. It is also more transparent, as the system focuses on prices, which are of greater concern to customers. With the creation of an independent regulatory body, the price-cap regulatory regime is less prone to political interference than was the case when utilities were nationalized, and the government interfered in the daily operations of the public utilities. A regulatory body, independent of government control, can make regulation more open and the rules of the game clearer than when the government is directly involved. Customers are guaranteed that price increases are under some form of control. Hence, the system helps to curb inflation and, being less discretionary, the price-cap is in less danger of regulatory capture.

Like all regulatory systems, the price-cap has its shortcomings. Many of the arguments put forward against price-capping revolve around the setting of the X factor. First, the X factor has to be set and reset repeatedly, in order to secure a reasonable rate of return for the regulated firm. In addition to production costs and the cost of capital, this process requires information on productivity gains. If the X factor is not set appropriately, allocative inefficiency will arise. In addition, there would be political pressure from the company and consumers in setting and resetting the X factor. Any shift in the X factor in favour of one group (customers) will be at the expense of the other (shareholders).

Vickers and Yarrow (1988) argued that price-cap formulas lack any long-term guarantees as to the decisions that will be made when they come to be reviewed. In the absence of clear guidance as to the long-term conduct of regulatory policy, private investors may be concerned that they will not be able to recover the cost of capital. Hence, because of the lack of credibility with respect to future government policy, there is a real danger of under-investment in a privatized industry. The cost of capital (and prices) may be raised by the presence of this regulatory risk.

Second, companies may believe that any increase in efficiency in the short term will invite a tougher X factor at a later stage, or even induce an adverse change in the X factor within the current period. If the short-term gain is more than offset by the long-term loss, an increase in productivity may be avoided. This is similar to the problem of the ratchet effect in the contract made between a socialist firm and the state.

Third, some people question whether price-cap actually has greater price flexibility and transparency. Under a rate-of-return regulation, the regulatory procedure, which involves public hearings and litigation, should be more transparent. Greater price flexibility may be a disadvantage rather than an advantage, since it allows price discrimination and cross-subsidization. As it is difficult to precisely determine the costs of various services, price discrimination and cross-subsidization cannot be easily detected.

Lastly, as a price-cap focuses on prices, regulated firms may fall behind on quality. Therefore, the regulator has to incur costs to closely monitor the quality of service.

## Concluding Remarks

At present, the two electricity companies are regulated by the Scheme of Control, while the towngas company is not. This situation may create a problem of unfair competition between power utilities in the energy industry. The government should avoid such asymmetry in its policies on power companies and provide a level playing field for all players within the energy industry. Although the two electricity companies have sealed their new schemes (which last until 2008) with the government, the new agreements contain a five-year review clause. This enables either party, the government or the regulated electric utility, to ask for a revision of the terms in the agreement. Such a request should be made before the end of the company's financial year in 1998. Hence, despite the fact that both companies are now operating under the new Scheme of Control, the Hong Kong government can still initiate changes in the terms of the scheme.

Over the past years, the operating environment of the energy industry has changed rapidly. It is time for the government in Hong Kong to review its existing policies and to seek ways of improving the existing control scheme. Hong Kong can borrow from the experiences of other countries to introduce structural and regulatory changes that will enhance competition within the energy industry.

# CHAPTER 5

# Enhancing Competition in the Hong Kong Energy Industry

## Competition in the Hong Kong Energy Industry

### Existing Competition

The fact that there are two electricity companies allows the government to compare and review the tariffs charged by them. The regulator can evaluate the potential of one company by using the production cost of the other. Each electricity company exerts pressure on the other to lower costs. This form of indirect yardstick competition provides a positive check on the tariff levels charged by the two companies. Of course, since the two companies are not operating in identical environments, it is difficult to discern to what extent the cost difference is due to efficiency in operation (Shleifer 1985). Apart from implicit yardstick competition, the two electricity companies compete for large users of electricity such as factory owners who can make a choice on their plant site to lower electricity costs. Before the early 1980s China Light and Power Company Limited (CLP) had a policy of "subsidizing industry". Bulk tariffs levels imposed on industrial users were much lower than those charged by Hongkong Electric Company Limited (HEC). This could partly explain why CLP had a much larger portion of industrial users.

In addition to indirect inter-regional competition, there is potential international competition among electricity companies operating in different countries. Although the two Hong Kong electricity companies are regional monopolists, prices charged by them cannot diverge greatly from prices elsewhere. The regulator and pressure groups may compare price differences between Hong Kong and other countries to judge whether the two electricity companies are operating efficiently. When comparing electricity prices across countries, one must be aware of the differences in the operating and regulatory environment between countries. Technologies and electricity systems used by companies in different countries are not the same, and the existence of hydroelectric power which is non-existent in Hong Kong — can greatly reduce electricity tariffs. In addition, government subsidies or taxes often distort price structure.

To sum up, inter-fuel competition and potential interregional and international competition have been restraining electricity and gas companies in Hong Kong from abusing their market power.

## Barriers to Competition

When introducing more competitive elements into the local energy industry, two major issues should be addressed. First, given Hong Kong's dense population and highly congested underground traffic, it may not be able to accommodate another gas or electricity transmission network. Besides, for safety and efficiency reasons, it is better to have a single network rather than multiple networks. Since the existing power companies have already developed their extensive transmission and distribution networks, a new supplier would find it difficult to enter the market, unless it could access existing networks. Dominant firms may attempt to exercise their power over the networks to restrict new entrants. This problem can be avoided by introducing an open access system (or a common carrier system) into the industry like that being done in the telecommunications industry.

Second, competition in the fuel market often starts at the planning stage of a new property development project. As the residential sector is the largest fuel consumer, competition is particularly keen in major real estate developments. Alternative fuel suppliers compete with one another to provide the supply system for cooking and water heating. They offer developers attractive terms for the right to install their supply systems. In some cases, they include appliances that are far below market cost as a further inducement in order that developers will adopt their supply systems. Both the gas and electricity companies are known to have used such strategies to win a greater market share at the planning stage of development projects.

If developers took the interest of the buyers of residential flats into consideration, they would choose a fuel supply system with the lowest cost. But structural impediments may prevent fair competition from taking place. Since the majority shareholders of the energy companies in Hong Kong are also majority shareholders in some property development firms, a fuel supply system may be chosen on the basis of serving the interests of these shareholders rather than the consumers' interests. For example, as HEC is controlled by Cheung Kong (Holdings) Limited, electric water-heaters may be selected over gas water-heaters in real estate development projects by Cheung Kong, even though the gas option may be more cost effective. Similarly, Hongkong and China Gas Company Limited (HKCG) is controlled by Henderson Land Development Company Limited, which may promote the use of towngas at the expense of other fuels in their development projects. Thus, this close business association between the energy companies and the property development business has been enabling energy producers to increase their market shares by preventing fair competition from taking place.

Access to transmission networks and opening up the market of big development projects are therefore pertinent ways to encouraging competition within the local energy industry.

## Improving Efficiency in the Electricity Industry

### Alternative Approaches to Improving Efficiency

When the government entered into negotiations with the two electricity companies concerning the terms of their new Schemes of

Control in the early 1990s, the public at large initiated several alternatives for the government to consider in order to protect customers and to promote efficiency within the electricity industry. These alternatives are briefly summarized below.

### *Merging the two companies*

This option see that the merging the two companies would lead to greater efficiency in the generation of electricity and cost-savings from larger economies of scale. The networks of the two companies have been interconnected since 1981 for the purpose of mutual backup facilities in the event of an emergency. Merging would also allow economic interchange of electricity, which would result in cost savings. It is expected that a full integration of the networks of the two companies would generate even larger cost savings. But the basic problem is the way in which the government can induce the two private companies to join forces. If integration would reduce costs and raise profits, then there is no economic reason why the two companies should not voluntarily decide to do so.

### *Deregulation*

Some interest groups have long argued for the scrapping of the Scheme of Control. They argue that the scheme has been protecting the electricity companies rather than the consumers. They feel that the two companies should not be guaranteed permitted rates of return. However, if the Scheme of Control were abandoned, what would the alternative be? One alternative would be a deregulation of the industry, which would allow for free competition in certain stages (e.g., generation) in the production of electricity. Both producers and consumers would benefit from market competition if the efficiency of the energy industry improved. The geographical boundaries in the supply of electricity could be removed. The interconnection of power systems would also make it possible for CLP and HEC to compete and to supply electricity to all the areas of Hong Kong.

### *Maintaining the status quo*

Both the government and the two companies argue that the existing system functions very well. Electricity prices in Hong Kong are low compared to those in other countries. Reliability of supply is high, with few major outages. The terms under the Scheme of Control, including the permitted rates of return, ensure sufficient investment in the industry. Sufficient investment is beneficial to the future growth of the economy.

After lengthy negotiations, the government eventually decided to renew the Scheme of Control for the two electricity companies. In broad terms, the new scheme remains more or less the same as its predecessor. However, the two companies and the government have agreed to increase the period over which fixed assets are depreciated in order to more accurately reflect the expected useful life of assets. Both companies are now obliged to promote demand-side management and energy conservation.

Once CLP and Exxon had signed the contract with the government, they announced their decision to spend up to HK$60 billion over the next decade on a new plant at Black Point. The first two units, with a total capacity of 625 MW, were commissioned in 1996. As in the previous scheme, CLP would rely on long-term loans (fixed interest rates ranging between 5% and 10.5%) to finance the project. In addition, the company has signed a 20-year long-term take-or-pay contract with China National Offshore Oil Corporation for the supply of natural gas from Hainan Island. However, the economic slowdown in Hong Kong and China since 1994, together with the continuing relocation of Hong Kong factories to China, have forced CLP to cut its capital expenditure from $60 billion to $52 billion and defer the commissioning dates of various generating and transmission projects.

After signing the contract for the new scheme, HEC was granted government approval to build two new 350-MW units on Lamma Island for the period 1994–1998. The company has also submitted a plan to construct a new power plant (with three 600-MW units) to meet the projected growth in electricity demand from

Hong Kong Island. The first 600-MW coal-fired unit will be commissioned in 2003. The entire project is expected to be completed by 2013.

## Improving the Scheme of Control

We support the original spirit of the Scheme of Control. Because of the nature of the electricity industry — which is characterized by, among other things, demand uncertainty, economies of scale, asset specificity, and a long planning horizon — a long-term contract protecting the producers' right to serve seems to be desirable. As producers' returns are protected to a certain extent, the Scheme of Control can introduce stability into the industry and facilitate debt financing. Financial institutions are then more willing to provide loans to the electricity companies and to finance the projects at a lower cost. This kind of investment protection is particularly useful for developing countries. For example, if China wants to attract foreign investment in building up its infrastructure, a control scheme with a development fund mechanism can be used to finance expansion, stabilize tariffs, and protect the interest of investors.

Our analysis of the Scheme of Control in Hong Kong, however, also suggests that, despite its early success, the Scheme has not been able to restrict the electricity companies' returns on equity capital to a level "which is reasonable in relation to the risks involved and the capital invested in and retained in the business". Unless by coincidence, a fixed maximum interest rate (8%) charged on debts and the development fund cannot reflect the cost of debt capital correctly. The distortion is even more serious during periods of high inflation and interest rates (as in the early 1980s). Because the permitted rate of return on debt is higher than the cost of debt capital, regulated utilities can effectively get around the regulation on asset return by altering their capital structures.

Empirical evidence has shown that the whole electricity industry can be regarded as a "natural monopoly": average cost decreases when output increases. However, this does not necessarily imply that the service should be provided by one single firm or that competition cannot exist in the industry. Competition is found to be feasible at the generation and supply stages. Although some analysts have proposed the merging of the two companies, we take the position that the two companies should remain separate entities, and that competition between them should be made possible. There are three ways to promote competitions.

### *Separating electricity supply from other businesses*

It has been found that the "beta value" of a regulated firm is influenced by its organizational form. A diversified electricity firm has a higher beta value than a non-diversified one that concentrates on the regulated business (Robison et al. 1995). Since the major objective of any regulatory contract is to protect both the producers' right to serve and consumers' right to be served, we should find ways in which we can measure precisely the cost of capital employed in the regulated industry. For the regulator to apply finance theories to measure capital cost, the two electricity companies should not be allowed to engage in businesses other than electricity supply. They could form separate companies for other businesses. Such a structural separation would prevent cost-shifting and cross-subsidization within the firm.

### *Introducing new regulatory arrangements*

Some interest groups have long urged the government to use shareholders' funds instead of fixed assets as the rate-base on which to calculate the permitted return, so as to prevent over-expansion by the utilities. But their argument does not address the major problem of the Scheme. Our analysis has shown that the two electricity companies have little incentive to use shareholders' funds to finance capital expansion, because the permitted rate of return on equity does not cover the equity cost, while the permitted rate on debt far exceeds the actual debt cost. If the permitted rate of return, either on equity capital or debt capital, is determined correctly to reflect their respective costs, then the regulated firm will be less inclined to increase the permitted return by changing its capital structure.

If fixed assets are maintained as the rate base, then the interest rate charged on debt and the development fund, instead of being fixed at a maximum of 8%, should be allowed to change in line with actual interest rate movements. Furthermore, the permitted return should be set in real terms, not in nominal terms. Instead of setting a fixed permitted rate on assets, the regulator should determine the equity cost, debt cost and inflation rate first, and then use the information to determine the overall cost of capital. The development fund could continue to serve as a return buffer for a regulated utility, as this reduces its risk. Moreover, the development fund also provides a regulated firm with some incentive to lower cost so as to accumulate the fund for future uses.

To encourage efficient production, a rate-of-return regulation could be imposed together with an arrangement similar to price-capping. Under existing arrangements, the two power companies have to submit their five-year financial plans, together with their tariff projections, to the government for approval. Once the government has granted approval for the plans, they can adjust tariffs according to their projections. If actual costs are below the forecast ones, they will be able to earn excess returns. But the excess returns earned will be transferred to the development fund, rather than retained by the companies. This may curtail their incentives to lower costs and prices.

To promote efficiency in production, we can modify the existing arrangements by allowing the two companies to share the benefits if they have managed to lower tariffs below the projected levels. The proposed new arrangement would not only allow the utilities to earn sufficient returns to cover their capital cost, but, more importantly, also encourage them to operate more efficiently within the regulatory period. After a five-year period, there should be a review to set new tariff levels for another five years. Based on the experience of the U.K., there is inevitably some political pressure on the regulator to tighten the projected tariff levels during the review period, when consumers realize that a utility has earned more than its permitted returns.

From the foregoing analysis, the government can build some incentives into the schemes when reviewing the financial plans submitted by the two electricity companies for the period 1998–2003 and 2003-2008. When the two electricity companies make projections on future demand and tariff levels, as they have done in previous reviews, the government can study their projections carefully. Based on the information on capital expenditure, productivity growth, and anticipated inflation and cost rises, it then determines the projected tariff levels. In *ex post*, if the two companies are able to set tariffs lower than these projected levels, they can share the benefit (in terms of higher returns) arising from lower tariffs with consumers instead of transferring all the surplus to the development fund.

### *Preventing over-expansion*

Under the existing Scheme of Control, the permitted return is entirely based on the fixed assets of the regulated company, regardless of whether the assets are "used and useful". The regulator has to exercise judgement when deciding to approve or disallow the expansion plan submitted by the regulated company. Once the plan has been approved, tariffs are adjusted accordingly to enable the company to earn the permitted return.

Electricity supply is a capital-intensive industry. The construction of generators and transmission networks requires huge capital investment. If the permitted rate is set too high, there will be a tendency for electricity companies to overestimate demand growth. If the permitted rate of return is set at a level to reflect the cost of capital correctly, then over-expansion can be curtailed. However, this lowers the regulated firm's incentive to reduce costs and shifts the risk from the producers entirely to the consumers. To prevent over-expansion, the government should design a mechanism to encourage truthful forecasts of demand growth. For example, the regulated firm should not be able to revise its projected tariffs upwards if actual demand falls short of its projections, or it has to share part of the financial burden that arises from inaccurate forecasts. In recent years, there has been an expansion of economic literature covering the design of self-revelation mechanisms.

## Enhancing Competition in the Electricity Industry

In the long term, if there is real competition within the electricity industry, then rate-of-return regulations or price controls imposed by the government would not be necessary. Studies conducted in the U.S. and the U.K. have found that an integrated transmission and distribution network can reduce costs substantially. Hence, a tentative approach would be to create a separate entity for transmission and distribution networks so that these networks are removed from the control of CLP and HEC. If such a structural separation proves to be difficult, we can maintain the existing industrial structure, but allow the two companies to share each other's networks by paying an access price. To achieve this, the regulator should enforce separation in the financial accounting of the generation and transmission businesses within the companies.

### Accounting Separation

Compared with a structural separation, an accounting separation can retain the benefits gained from the economies of scope and economies of scale, which are important features for a vertically integrated firm in the electricity industry. The major task facing the regulator under a system of accounting separation is to discern whether the regulated firm has manipulated the cost data and has allocated costs to its own advantage. When the two private companies fail to reach a voluntary agreement, the government should act as an arbitrator in setting the access price that permits one company to use the transmission network of the other. In Hong Kong, a similar arrangement has already been made to allow the three new telephone companies' access to Hongkong Telecom's (HKT) network. In determining what a "fair" access price should be, the government can continue to rely on a system of rate-of-return regulation.

## Competition in Generation and Supply

The government could change the existing regional monopoly by allowing competition in certain stages of electricity production and supply. On the generation side, in which economies of scale are not obvious, HEC should be allowed to compete with the associated generating companies of CLP in Hong Kong and China. If the purchase price of electricity charged by the other utility is lower than a utility's own generation cost, the utility should be mandated to purchase electricity. Instead of building new power plants in Hong Kong, CLP and HEC should also be encouraged to purchase cheaper energy from areas outside Hong Kong. To ensure reliability in supply, these companies can rely on long-term power supply contracts. The case of the U.S. in the 1980s is illuminating in this respect. The companies should purchase electricity from other areas if the purchase price is lower than the "full avoided" cost explained below.

Full avoided cost is defined as the incremental cost of generating electricity (including capacity cost) incurred by an electricity company. An arrangement of mandatory purchase (in the preceding paragraph) would enhance competition, put pressure on the two electricity companies to lower costs, and close their existing tariff differential.

The experience of American utilities in recent years is useful here. The reality of an increasingly stringent prudence review, together with the political and environmental impediments to the construction of large centralized generating facilities, has encouraged U.S. utilities to concentrate on their transmission and distribution functions and to look elsewhere for their power requirements (Crocker and Masten 1996).

Experience gained from the U.K. shows that competition in power generation is not workable if the market is dominated by one or two large firms. To decrease the market power exercised by CLP under a duopolistic market structure, new entrants into the energy generation market should be encouraged. In this respect, encouraging the import of electricity from China may also help improve the situation. The completion of the Daya Bay nuclear power station and other generating units in Southern China has resulted in a surplus of electricity in certain areas of the Guangdong Province. Instead of building new generators in Hong Kong, we could make use of the surplus energy in China by encouraging private companies to import electricity through long-term contracts. Apart from removing many of the environmental problems associated with power generation, this would increase the number of competitors in our electricity industry. This option is feasible because the electricity systems in Hong Kong and Guangdong are already interconnected.

On the supply side, both companies can go beyond their existing transmission networks to supply electricity to new customers. They can interconnect with another company upon paying a fee. As a first step, they can make long-term contracts with large users (e.g., industrial and commercial users, and all users in the same housing estates) and then plan their capacity expansion ahead of time. Demand for electricity on Lantau Island, in the western New Territories, in western Kowloon, and in many other newly developed areas has increased because of the construction of the new airport and other developments in Hong Kong. The government can utilize this opportunity to introduce competition into the electricity supply industry either at the wholesale or the retail level. In the U.S. and the U.K. competition at the wholesale level and in co-ordination sales has been found to be feasible and has enhanced efficiency in the industry. Full competition at the retail level will also commence in the near future. However, since the two companies' systems are interconnected via a tie line with a 720-MVA capacity, competition between them is possible.

Under the existing Scheme of Control, the two electricity companies are obliged to promote demand-side management and energy conservation. Since the permitted return is entirely based on fixed assets, they have little incentive to promote conservation. The reason is that their asset base and permitted return would be reduced if they were able to successfully control demand growth. Allowing competition in the electricity industry would provide an electricity company with an incentive to promote demand-side management. A firm under competition will try to adopt different pricing arrangements to lower the prices paid by the customers. By encouraging users to shift their demands to off-peak periods, the firm can effectively reduce the costs and prices of supplying electricity.

The government can also consider an option of using utility avoided cost to encourage energy conservation. At present, the capacity expansion of any power company is for the purpose of meeting projected peak demand. If an electricity company can successfully introduce measures to reduce growth in peak demand, some operating and capital expenditures can be avoided. To encourage energy savings, the government can design an incentive system that allows both the utility and the customers to share the expenditures avoided through conservation efforts (Reddy 1996).

After the government has fostered a programme to promote competition in the generation and supply of energy, a Scheme of Control is then only applied to the transmission and distribution sides of the business. The regulator's major task is to facilitate the interconnection process and the determination of access charges. The regulator will only step in if the regulated firms are found to be adopting anti-competitive strategies to deter new entrants. At present, in a comparable case, the regulator in Hong Kong's telecommunications industry has allowed a network provider to earn a fair return on its assets when determining the interconnection fees.

## Restructuring the Hong Kong Gas Industry

The restructuring of the electricity industry discussed above should also be applied to the gas industry. Because of the slowing demand for electricity in Hong Kong and from China, CLP is now running with a substantial surplus in its generating capacity. CLP should therefore be required to delay its plan to build gas-fired generators. The natural gas saved from generating electricity could then be used for other purposes, such as gas heating and cooking. This would then put competitive pressure on the towngas company to improve efficiency. In addition, because of the desirable characteristics of natural gas (e.g., it has higher thermal efficiency and is cleaner and cheaper), the government should encourage its uses in other sectors. However, unless the government intervenes, it is not in the commercial interest of CLP to further delay its capacity expansion and to divert natural gas from electricity generation to domestic use.

A study of relevant experiences in the U.K. and the U.S. is essential to the restructuring of Hong Kong's gas industry. On the one hand, we need to avoid high transaction costs faced by gas users as a result of the unbundling of various services; while on the other hand, we need to prevent a dominant firm from exercising its market power over the network to shut out competition.

## Separation of Transportation from Production and Supply

Studies have indicated that the production and supply of gas is not naturally monopolistic and should be separated from the transportation function. Instead of encouraging competition from alternative fuels, we need to emphasize gas-to-gas competition. Hence, a pro-competition measure would require HKCG to separate its production, as well as its supply business, from transportation. New suppliers of gas should be able to gain equal access to HKCG's pipelines, contract with gas producers (other than HKCG's production plants) and request HKCG's transportation services upon paying a non-discriminatory price.

Since HKCG is now running with significant excess capacity, one may argue that production units installed by a new competitor may not be socially desirable (Kwong 1995). Also, as HKCG has an installed pipeline network of almost 2,000 kilometres, no entrant can hope to build a second network. It will be difficult for a new entrant to find a scale of operation that would be able to undercut HKCG. But the introduction of natural gas provides a possibility for competition. New natural gas suppliers could make use of HKCG's network to distribute natural gas. The existing towngas network can be used for natural gas transportation without any modification. Alternatively, new gas suppliers can elect to construct their own networks and then interconnect with HKCG's network. The regulator's role is to prevent HKCG from exercising its market power over the network to foreclose competition. The shift from towngas to natural gas will only require gas users to modify their appliances with a moderate expense.

Competition in the production and supply of gas would lower the prices paid by end users. If natural gas is cheaper, or if HKCG's existing production plants are inefficient, new firms (including CLP) would be able to easily enter the market and compete with HKCG's production business. Since the production cost of natural gas rises as supplies in gas field dwindle, and since the transportation cost from the gas field is higher, natural gas may not be cheaper than towngas in the long run. There have been reports that Sun Hung Kai Industries is negotiating with CLP to buy its excess natural gas, which would be sold to industrial users. But CLP's surplus supply is limited; further supply of natural gas to Hong Kong depends on new discoveries in the region. Another option is to import liquefied natural gas (LNG) from other countries. This will again depend on private sector initiative, and safety measures are also required in the transportation of LNG.

## Scheme of Control on Transmission

As was discussed above, competition in the gas production and supply market requires HKCG to separate its transmission business from other businesses. The government should decide carefully whether HKCG's transmission business should be subject to rate-of-return regulation similar to the Scheme of Control. The purpose of the rate-of-return control is to avoid unjust confiscation of private property. In addition, for an open access or a common carrier system to be viable, the regulator may have to determine the appropriate method for restructuring HKCG. Feasible alternatives like vertical separation or accounting separation should be considered carefully.

When applying the U.S. and U.K. models of restructuring and regulation to the Hong Kong gas industry, we have to take into account our specific local circumstances. While the U.S. and the U.K. have extensive production, transmission, and distribution systems based on natural gas, Hong Kong is likely to continue relying on a large local towngas distribution system in the foreseeable future. Although we can borrow from the U.S. and U.K. experiences of separating the gas transportation function from distribution, experiences in other countries, which depend on a local transmission and distribution network, might be more applicable.

An interesting comparison is the Japanese city gas distribution systems in Tokyo and Osaka. Both systems are extensive but with little or no high pressure transmission and no local natural gas production. In March 1995 the gas industry in these two cities was liberalized, and large customers (over 2 million cubic metres per year) are now allowed to choose their own local distributor. It is expected that this will put pressure on the smaller and less efficient towngas distribution companies as well as on LPG suppliers. We can borrow from the Japanese experience by first liberalizing a certain market segment of the supply business.

In order to achieve a smooth transition from a monopolistic market to a competitive one, the Hong Kong government may decide to open certain market segments first. The construction of the new airport and other new developments have opened up many new areas that require fuel supply. Alternative fuel providers, which include suppliers of electricity, towngas, natural gas, and LPG, can bid for the supply contracts on constructing the energy systems in these newly developing areas. If competition in these areas proves to be successful, then it can be extended to other areas. But unless the government has provided a "level playing field", energy companies may resist liberalization and avoid direct competition with one other.

## A Unified Regulatory Body

To enhance competition in the telecommunications industry in Hong Kong, the Office of the Telecommunications Authority

(OFTA) was established in July 1993. Since its inception, OFTA has taken over all the tasks previously performed by the telecommunications branch of the government post office. In the energy industry, different government departments (which include the Economic Services Branch, the Electrical and Mechanical Services Department, the Gas Standards Office, the Environmental and Protection Department, among others) share the responsibility of regulating the power companies in Hong Kong. Consequently, the government's energy policies tend to be piecemeal. As the energy system in Hong Kong expands and integrates with the system in Southern China, the government should consider whether it is necessary to set up a unified regulatory body similar to OFTA to govern the whole energy industry.

It is expected that the government will deal with more and more issues relating to regulation and competition in the energy industry. These issues include the planning and construction of new power stations, energy safety and conservation, demand-side management, environmental problems associated with power generation and transmission, market competition in the fuel market, and power purchase and the interconnection between energy companies within and outside Hong Kong. At present, government officials charged with the responsibility of regulating the energy industry may lack a clear understanding of the whole energy industry. There is also a danger of political influence and regulatory capture in the process of formulating energy policies. An independent, unified regulatory body covering all areas of the energy industry can better formulate consistent regulatory and competitive policies.

In 1995 a member of the Legislative Council submitted a complaint of maladministration against the Economic Services Branch to the Commissioner for Administrative Complaints. The complaint was related to the massive over-capacity of the two electricity companies, which had raised electricity charges in Hong Kong. The legislator also complained about the lack of transparency in the monitoring process of the Branch. The Economic Services Branch responded by arguing that the expansion plans had been assessed by independent consultants and approved by the

Executive Council. There were also other government departments involved in the monitoring process, and the Branch should not be held solely responsible for the over-capacity of the two electricity companies. However, the complaint case was dismissed later because since the Commissioner for Administrative Complaints had *no jurisdiction* over the Executive Council which approved the Scheme of Control and the expansion plans of the two electricity companies.

Although the existing monitoring arrangement can provide checks and balances between various government bodies, the above example vividly illustrates the deficiencies which exist in the system: a lack of accountability and transparency. As the two regulated electricity companies are required to make detailed information available to the public, the unregulated towngas company may make use of this information to increase its market power. A separate and independent body may be able to improve the regulatory process. All the three power companies should be subject to the same regulatory control. The proposed unified regulatory body should be held responsible for all the policy decisions it makes, and be directly accountable to the Executive Council and the public. The Commissioner for Administration Complaints should be allowed to receive complaints and to investigate the operations of this regulatory body.

In its "competition report" (1995), the Consumer Council also proposed the establishment of an Energy Commission to co-ordinate all energy issues within the fuel market. The Commission would be advised by an Energy Advisory Committee. It should be emphasized that the role of this regulatory body, if created, is restricted to formulating coherent policies to improve efficiency and enhance competition within the energy industry.

When introducing changes into the existing system, there are inevitably some trade-offs. Increased competition in the electricity industry would eliminate cross-subsidization among different groups of users. There is also a problem of co-ordinating different energy systems and ensuring system reliability. Hence, competition cannot totally replace co-operation among utility companies.

Furthermore, open competition would raise a firm's business risk and the cost of capital. We have to weigh the benefits of increased efficiency against the higher cost of capital incurred.

# CHAPTER 6

# Recommendations and Conclusion

## Recommendations

In recent years there have been clear trends in the energy industry towards increased competition. The governments of many Organisation for Economic Cooperation and Development (OECD) countries have tried to restructure their utility sectors or to institute regulatory reforms that include competitive elements. Restructuring in the electricity and gas industries is underway in many of these countries where free competition in production and supply and open transmission access have gradually replaced vertically integrated monopolists. With an increased number of players in the energy industry, the governments have tried to pay particular attention to security of supply. An approach of increased co-operation and competition has been emphasized in the restructuring process.

With regards to the energy industry in Hong Kong, the market is still dominated by three big players. Direct competition between the two electricity companies does not exist, and competition between towngas and other fuel (like electricity and liquefied petroleum gas suppliers) is also limited. Although the tariffs charged by these utility companies are still held within reasonable limits, evidence has shown that there is room for improving efficiency. It is believed that increased competition in the market can help prevent over-expansion, improve operating efficiency, and lower energy costs. We summarize our approach in four points on how to enhance competition in the Hong Kong energy industry — Compete in production and supply, control the transmission business, promote efficiency, and unify the regulatory body.

## Increase Competition in Production and Supply

In order to change the existing situation of regional monopoly, the government could allow competition in certain stages of electricity production and supply. On the generation business in which scale economies are not obvious, Hongkong Electric Company Limited (HEC) should be allowed to compete with the associated generating companies of China Light and Power Company Limited (CLP) in Hong Kong and China. If the purchase price of electricity charged by the other utility is lower than a utility's own generation cost, the utility should be mandated to purchase electricity. Instead of building new power plants in Hong Kong, CLP and HEC should be encouraged to purchase cheaper energy from areas outside Hong Kong; they should also be encouraged to rely on long-term power supply contracts in order to ensure reliability in supply.

Concerning the supply business, both companies could go beyond their existing transmission networks to supply electricity to new customers. They could inter-connect with the other company upon paying a fee. As a first step, they could make long-term contracts with large users (e.g., industrial and commercial users and users in the same housing estates) and then plan their capacity expansion. The construction of the new airport and other developments in Hong Kong have increased the demand for electricity on Lantau Island. Demand is also rising in the western side of the New Territories, western Kowloon and other newly developed areas. The government should utilize this opportunity to introduce competition at the wholesale (or even the retail) level into the electricity supply industry.

Studies have indicated that production and supply of gas is not naturally monopolistic, so production and supply should be separated from the transportation function. Instead of encouraging competition from alternative fuels, we need to emphasize gas-to-gas competition. Hence, a new structure for enhancing competition in the gas industry may require Hongkong and China Gas Company Limited (HKCG) to separate its production, as well as its supply business, from transportation. New suppliers of gas (including natural gas) should be able to obtain equal access to HKCG's pipelines, to contract with gas producers (other than HKCG's production plants) and to request use of HKCG's transportation services upon paying a non-discriminatory price.

In order to smooth the transition from a monopolistic market to a competitive market, the government may decide to open some market segments first. Alternative fuel providers, which include suppliers of electricity, towngas, natural gas and LPG, can bid for supply contracts relating to the construction of the energy systems in the newly developing areas in Lantau and northwest New Territories. If competition in these areas proves to be successful, it can be extended to other areas.

## Scheme of Control Imposed on Transmission Business

After production and supply businesses are opened to competition, the Scheme of Control can then be narrowed to the transmission and distribution businesses. In the electricity industry, a tentative approach would be to put the transmission and distribution networks under a separate entity other than CLP and HEC. If such a structural separation proves to be difficult, the two companies would be allowed to share each other's networks by paying an access price. To achieve this, the regulator should enforce accounting separation between the generation and transmission businesses of the two electricity companies. The major task of the regulator would be to facilitate the inter-connection process and the determination of access charges. The regulator would step in only if it is found that the regulated firms are adopting anti-competitive strategies to deter new entrants.

Competition in the gas production and supply market would also require HKCG to separate its transmission business from other businesses. The government should decide carefully whether HKCG's transmission business should be subject to a rate-of-return regulation similar to that of the Scheme of Control. The purpose of the rate-of-return control has been to avoid unjust confiscation of private property. In addition, for open access or a common carrier system to be viable, the regulator might have to determine the appropriate method for restructuring HKCG. Feasible alternatives such as vertical separation or accounting separation should be considered seriously.

When imposing a Scheme of Control on the transmission business of power companies, some refinements would be needed. If fixed assets are to maintain as the base rate, then the interest rate charged on debts and the development fund, instead of being fixed at a maximum of 8%, should be allowed to change in line with movements in the actual interest rate. Furthermore, the permitted return should be set in real terms, not in nominal terms. Instead of setting a fixed permitted rate on assets, the regulator should first determine the equity cost, debt cost and inflation rate, and then use the information to determine the overall cost of capital. The development fund could continue to serve as a return buffer for a regulated utility, as this reduces the risk faced by a regulated utility. Moreover, the development fund would also provide some incentives for a regulated firm to lower costs in order to accumulate the fund for future uses.

To encourage efficient production, the rate-of-return regulation could be imposed together with some arrangement similar to price-capping. Under the existing arrangements, the two electricity companies have to submit their five-year financial plans, together with their tariff projections, to the government for approval. Once the government approves the plan, the companies may adjust tariffs according to their projected forecasts. To induce efficiency in production, the existing arrangements might be modified by allowing the two companies to share the benefits if they manage to lower transmission charges below the projected levels. The proposed new arrangements would not only allow the utilities to earn sufficient returns to cover its cost of capital, but, more importantly, also encourage the utilities to operate more efficiently within the regulatory period. After five years, there would be a review to set new transmission charges for another five-year period.

## Promoting Energy Efficiency

The restructuring of the electricity industry should involve the gas industry. Because of the slowing demand growth for electricity in both Hong Kong and China, CLP is now running with substantial surplus in its generating capacity. CLP has delayed its expansion plan to build gas-fired generators. The natural gas saved from generating electricity could instead be used for other purposes such as gas heating and cooking. This would put competitive pressure on the towngas company to improve efficiency. In addition, because of the desirable characteristics of natural gas (e.g. efficient, clean and cheap), the government should encourage its use in other sectors. It is expected that unless the government makes the request, it is not in the commercial interest of CLP to further delay its capacity expansion and to divert the use of natural gas.

Competition in the production and supply of gas would lower the prices paid by end users. If natural gas is cheaper, or if HKCG's existing production plants are inefficient, new entrants (including CLP) would easily enter the market and compete with HKCG's production business. But as CLP's surplus of gas is limited, further supplies of natural gas to Hong Kong will depend on new discoveries in the region. Another option is to encourage private firms to import liquefied natural gas (LNG) from other countries. However, it is important to note here that extra safety measures are required in the transportation of LNG.

Electricity demand has a summer peak, while gas demand has a winter peak. In summer, there is a greater electricity demand for air-conditioning. The two electricity companies have to increase installed capacity to meet the summer peak. In winter, there is an increased demand for gas used for heating and cooking. If the gas companies in Hong Kong successfully promote their businesses and lower Hong Kong's demand for electricity during summer, then the

**Figure 6.1**
**Proposed New Structure of the Energy Industry in Hong Kong**

| **Sources of Production** | | | |
|---|---|---|---|
| Electricity | | vs | Gas |
| CLP | HEC | Towngas | Natural gas |
| (Competition between companies for large users) | | (Competition in new development sites) | |

| **Scheme of Control** | | |
|---|---|---|
| Electricity | vs | Gas |
| Transmission & Distribution (Regulated by Government) | | Transmission & Distribution (Regulated by Government) |

| **Sources of Demand** | | |
|---|---|---|
| Electricity | vs | Gas |
| Competition in Marketing and Sales to Energy Users | | |

Note: CLP is China Light and Power Company Limited
HEC is Hongkong Electric Company Limited

pressure for the installation of new electricity generating facilities would be reduced. Consequently, overall energy costs in Hong Kong would decrease.

Allowing competition in the electricity industry would provide an incentive for an electricity company to promote demand-side management. A firm under competition would try to adopt different pricing arrangements to lower the prices paid by the customers. By encouraging users to shift their demands to off-peak periods, a firm could effectively reduce the costs and prices of supplying electricity. The government could also consider an option of using "utility avoided cost" to encourage energy conservation. To encourage energy savings, the government could design some incentive systems for the utilities and the customers to share in the avoided expenditure which would result from their conservation efforts.

Figure 6.1 shows the proposed new structure of the energy industry in Hong Kong. Direct competition should be introduced in the stages of energy production and supply (top panel in Figure 6.1). The two electricity companies should not compete only with each other, but also with other generators in China. They should be allowed to supply electricity beyond their existing servicing areas. At the supply stage, HKCG has to face competition not only from other LPG suppliers but also from potential natural gas suppliers which may include CLP.

A system of open access should be introduced in the transmission stage (middle panel). The rate of return realized by the transmission monopolists could be regulated by the Scheme of Control, yet the transmission capacity (other than firm capacity) could still be distributed among users by competition, such as the use of an open bidding system. The goal is to promote competition between suppliers of various forms of energy at the sources of demand (bottom panel).

## A Unified Regulatory Body

In addition to the importation of coal and oil products from China, Hong Kong is now importing electricity and natural gas from China. As the energy system in Hong Kong expands and integrates with the system in Southern China, it is expected that the government will deal with more and more issues relating to the regulation and competition in the energy industry. In view of this development, the government should consider whether it would be necessary to set up a separate, unified regulatory body to govern the whole energy industry in South China. It is hoped that an independent body could increase the transparency and accountability of the regulatory process. All the three power companies should be subject to the same regulatory control. The two roles of this regulatory body, if created, should be restricted to formulating coherent rather than piecemeal policies, and to enhance efficiency and market competition in the energy industry.

## Conclusion

Despite recent regulatory reforms in the telecommunications and bus industries, the Hong Kong government has been hesitant about reforming the energy industry. The two electricity companies will remain protected by the Scheme of Control until 2008. Undoubtedly, a long-term contract between a private firm and the government, such as the Scheme of Control, has its own benefits. But as evidence has shown, the Scheme fails to contain price increases because the returns on utilities are guaranteed, and the permitted rates have not been determined appropriately. The government has also failed to design any concrete policy to prevent the towngas company from abusing its market power in the fuel market. In sum, there is a need to introduce policies which can inject incentives and competitive forces into the energy industry.

One important lesson to be learned from the U.S. and U.K. experiences is that structural reforms are pertinent to promoting competition in the energy industry. Problems encountered in the U.K. after privatization could have been avoided had the energy industry been restructured in such a way that the dominant firms' foreseeable market power can be eliminated. Apart from encouraging new entrants and imposing an open access requirement on the transmission companies, the determination of access terms is also of paramount importance for real competition to take place. Hong Kong is fortunate to have learnt from other countries' experiences, and it need not proceed on a trial-and-error basis when Hong Kong introduces competition into the energy industry.

# Appendix

**Table 3.1**
**Average Electricity Prices in Hong Kong, 1948–1995**

| Average price (cents/kWh) | | | | Average price (cents/kWh) | | | |
|---|---|---|---|---|---|---|---|
| Year | CLP(1) | CLP(2) | HEC | Year | CLP(1) | CLP(2) | HEC |
| 1948 | 21.90 | | 20.78 | 1972 | 12.72 | | 14.06 |
| 1949 | 18.71 | | 20.10 | 1973 | 11.97 | | 13.63 |
| 1950 | 17.08 | | 18.59 | 1974 | 17.18 | | 20.74 |
| 1951 | 15.77 | | 17.18 | 1975 | 20.58 | | 22.34 |
| 1952 | 18.25 | | 19.36 | 1976 | 20.83 | | 22.12 |
| 1953 | 17.56 | | 18.55 | 1977 | 20.97 | | 22.90 |
| 1954 | 16.38 | | 17.61 | 1978 | 21.07 | | 22.80 |
| 1955 | 15.76 | | 17.21 | 1979 | 23.51 | | 28.40 |
| 1956 | 15.51 | | 17.15 | 1980 | 34.06 | | 38.10 |
| 1957 | 16.16 | | 17.74 | 1981 | 49.54 | | 55.50 |
| 1958 | 16.00 | | 17.33 | 1982 | 52.60 | | 55.60 |
| 1959 | 15.70 | | 16.80 | 1983 | 54.68 | | 60.60 |
| 1960 | 15.40 | | 16.36 | 1984 | 55.27 | | 60.50 |
| 1961 | 14.90 | | 16.07 | 1985 | 55.09 | 57.54 | 60.20 |
| 1962 | 14.72 | | 15.95 | 1986 | 54.27 | 56.10 | 58.00 |
| 1963 | 14.46 | | 15.52 | 1987 | 53.71 | 53.40 | 56.40 |
| 1964 | 14.31 | | 14.93 | 1988 | 53.32 | 53.23 | 56.20 |
| 1965 | 14.06 | | 14.35 | 1989 | 54.13 | 53.16 | 58.60 |
| 1966 | 13.54 | | 14.23 | 1990 | 55.49 | 55.66 | 63.40 |
| 1967 | 13.15 | | 14.23 | 1991 | 55.78 | 57.29 | 67.70 |
| 1968 | 13.08 | | 14.11 | 1992 | 56.79 | 59.28 | 71.10 |
| 1969 | 12.78 | | 13.75 | 1993 | 56.83 | 61.75 | 74.00 |
| 1970 | 12.00 | | 13.83 | 1994 | 66.74 | 64.64 | 77.50 |
| 1971 | 12.66 | | 14.02 | 1995 | 72.39 | 71.85 | 80.40 |

Source: Annual reports of CLP and HEC.
Note: For CLP(1): the average price is obtained by dividing the sales revenue by the quantities sold.
For CLP(2) and HEC: the average price is adjusted for fuel cost change and other rebates.

**Table 3.2**

**Inflation Rates and Electricity Prices (1973=100) in Hong Kong, 1973–1995**

| Period | CLP | CPI | Period | HEC | CPI |
|---|---|---|---|---|---|
| 10/72-9/73 | 100 | 100 | 1/73-12/73 | 100 | 100 |
| 10/73-9/74 | 144 | 117 | 1/74-12/74 | 152 | 114 |
| 10/74-9/75 | 172 | 123 | 1/75-12/75 | 164 | 116 |
| 10/75-9/76 | 174 | 127 | 1/76-12/76 | 162 | 120 |
| 10/76-9/77 | 175 | 133 | 1/77-12/77 | 168 | 127 |
| 10/77-9/78 | 176 | 140 | 1/78-12/78 | 167 | 134 |
| 10/78-9/79 | 196 | 154 | 1/79-12/79 | 208 | 150 |
| 10/79-9/80 | 285 | 179 | 1/80-12/80 | 280 | 173 |
| 10/80-9/81 | 414 | 203 | 1/81-12/81 | 407 | 200 |
| 10/81-9/82 | 439 | 221 | 1/82-12/82 | 408 | 220 |
| 10/82-9/83 | 457 | 249 | 1/83-12/83 | 445 | 243 |
| 10/83-9/84 | 462 | 274 | 1/84-12/84 | 444 | 262 |
| 10/84-9/85 | 460 | 283 | 1/85-12/85 | 442 | 271 |
| 10/85-9/86 | 453 | 291 | 1/86-12/86 | 426 | 278 |
| 10/86-9/87 | 449 | 305 | 1/87-12/87 | 414 | 293 |
| 10/87-9/88 | 445 | 327 | 1/88-12/88 | 412 | 315 |
| 10/88-9/89 | 452 | 358 | 1/89-12/89 | 430 | 347 |
| 10/89-9/90 | 464 | 392 | 1/90-12/90 | 465 | 381 |
| 10/90-9/91 | 466 | 438 | 1/91-12/91 | 497 | 427 |
| 10/91-9/92 | 474 | 471 | 1/92-12/92 | 522 | 467 |
| 10/92-9/93 | 475 | 512 | 1/93-12/93 | 543 | 507 |
| 10/93-9/94 | 558 | 554 | 1/94-12/94 | 569 | 548 |
| 10/94-9/95 | 605 | 603 | 1/95-12/95 | 590 | 596 |

Note: The financial period of CLP is from October to September next year.

**Table 3.3**
**Revenue Requirements of CLP and HEC, 1979–1995**

**(1) Percentage shares of different costs in total revenue (1979–1995)**

| | CLP | | | HEC | | |
|---|---|---|---|---|---|---|
| Year | Fuel cost (%) | Other costs (%) | Permitted return (%) | Fuel cost (%) | Other costs (%) | Permitted return (%) |
| 1979 | 56 | 23 | 20 | 47 | 26 | 27 |
| 1980 | 62 | 16 | 21 | 54 | 20 | 26 |
| 1981 | 67 | 12 | 22 | 59 | 15 | 26 |
| 1982 | 57 | 15 | 28 | 43 | 23 | 35 |
| 1983 | 49 | 20 | 31 | 31 | 37 | 32 |
| 1984 | 44 | 22 | 34 | 26 | 40 | 34 |
| 1985 | 37 | 25 | 38 | 26 | 37 | 37 |
| 1986 | 26 | 34 | 40 | 21 | 38 | 40 |
| 1987 | 20 | 39 | 40 | 17 | 44 | 40 |
| 1988 | 19 | 42 | 40 | 15 | 45 | 40 |
| 1989 | 20 | 42 | 38 | 19 | 39 | 42 |
| 1990 | 21 | 41 | 38 | 18 | 38 | 44 |
| 1991 | 23 | 39 | 38 | 18 | 38 | 45 |
| 1992 | 22 | 42 | 36 | 16 | 38 | 46 |
| 1993 | 21 | 43 | 36 | 14 | 40 | 46 |
| 1994 | 13 | 48 | 40 | 13 | 40 | 47 |
| 1995 | 12 | 47 | 40 | 13 | 33 | 53 |

**(2) Cost shares in average price (1993)**

| | CLP | HEC |
|---|---|---|
| Average price (cents/kWh) | 61.8 | 74.0 |
| Fuel cost | 12.0 (19.4%) | 11.5 (15.5%) |
| Other costs | 29.9 (48.4%) | 30.8 (41.6%) |
| Net return | 19.9 (32.2%) | 31.7 (42.8%) |

Source: Annual reports of CLP and HEC.

**Table 3.4**
**Real Electricity Prices (1973=100) in Asian Countries, 1973–1990**

| Period | Residential prices | | | Average prices | |
|---|---|---|---|---|---|
| | India | Indonesia | South Korea | CLP | HEC |
| 1973 | 100 | 100 | 100 | 100 | 100 |
| 1974 | 107 | 99 | 111 | 123 | 133 |
| 1975 | 124 | 117 | 145 | 140 | 142 |
| 1976 | 127 | 127 | 136 | 137 | 136 |
| 1977 | 138 | 116 | 132 | 131 | 133 |
| 1978 | 142 | 103 | 112 | 125 | 125 |
| 1979 | 132 | 77 | 132 | 127 | 139 |
| 1980 | 126 | 88 | 168 | 159 | 162 |
| 1981 | 132 | 83 | 182 | 204 | 204 |
| 1982 | 144 | 100 | 184 | 198 | 185 |
| 1983 | 143 | 119 | 171 | 184 | 183 |
| 1984 | 144 | 140 | 164 | 169 | 169 |
| 1985 | 119 | 132 | 159 | 163 | 163 |
| 1986 | 116 | 127 | 148 | 156 | 153 |
| 1987 | 115 | 108 | 138 | 147 | 141 |
| 1988 | 119 | 101 | 123 | 136 | 131 |
| 1989 | 118 | 114 | 118 | 126 | 124 |
| 1990 | 117 | 102 | 110 | 118 | 122 |

Source: *Energy in Developing Countries* (1994).

**Table 3.5**
**Average Electricity Prices and Equity Returns in Asian Countries**

| Countries | January 1993 | | January 1996 | | 1993/94 |
|---|---|---|---|---|---|
| | Price | Index | Price | Index | Equity return |
| Indonesia | 51 | 90 | 53 | 73 | 4% |
| Thailand | 56 | 98 | 59 | 81 | 23% |
| Korea | 64 | 114 | 66 | 90 | 8% |
| Singapore | 55 | 96 | 64 | 87 | 8% |
| Taiwan | 71 | 125 | 68 | 93 | 10% |
| Malaysia | 55 | 97 | 64 | 88 | 14% |
| Hong Kong | 57 | 100 | 73 | 100 | 20% |
| Philippines | 78 | 137 | 87 | 120 | 15% |
| Japan | 116 | 203 | 141 | 193 | 3% |

Source: *The Straits Times*, 12 February 1996 and 18 March 1996.
Note: Prices are in cents per kWh; index is 100 for the price in Hong Kong.

**Table 3.6**
**Debt-Equity Ratios and Beta Values, 1979–1992**

| Period | CLP | | | HEC | | |
|---|---|---|---|---|---|---|
| | D/E | T/E | Beta | D/E | T/E | Beta |
| 1978 – 82 | 0.71 | 0.07 | 0.97 | 0.23 | 0 | 0.94 |
| 1979 – 83 | 1.02 | 0.06 | 0.97 | 0.24 | 0.01 | 0.87 |
| 1980 – 84 | 1.25 | 0.06 | 0.94 | 0.24 | 0.02 | 0.76 |
| 1981 – 85 | 1.42 | 0.06 | 0.96 | 0.29 | 0.03 | 0.80 |
| 1982 – 86 | 1.43 | 0.08 | 0.96 | 0.30 | 0.04 | 0.77 |
| 1983 – 87 | 1.38 | 0.10 | 0.88 | 0.31 | 0.05 | 0.72 |
| 1984 – 88 | 1.22 | 0.12 | 0.91 | 0.34 | 0.06 | 0.70 |
| 1985 – 89 | 1.06 | 0.14 | 0.94 | 0.40 | 0.06 | 0.69 |
| 1986 – 90 | 0.85 | 0.16 | 0.95 | 0.39 | 0.06 | 0.65 |
| 1987 – 91 | 0.72 | 0.17 | 0.93 | 0.44 | 0.06 | 0.63 |
| 1988 – 92 | 0.63 | 0.18 | 0.92 | 0.48 | 0.06 | 0.62 |

**Table 3.7**
**Consumption of Towngas and LPG (in TJ), 1982–1991**

| Year | Towngas | LPG | Total |
|---|---|---|---|
| 1982 | 4,858 (46) | 5,739 (54) | 10,597 |
| 1983 | 5,924 (49) | 6,208 (51) | 12,132 |
| 1984 | 6,907 (51) | 6,683 (49) | 13,590 |
| 1985 | 7,979 (54) | 6,876 (46) | 14,855 |
| 1986 | 9,043 (55) | 7,251 (45) | 16,294 |
| 1987 | 10,584 (57) | 7,930 (43) | 18,514 |
| 1988 | 12,247 (59) | 8,347 (41) | 20,594 |
| 1989 | 13,671 (61) | 8,608 (39) | 22,279 |
| 1990 | 15,056 (62) | 9,138 (38) | 24,194 |
| 1991 | 16,238 (64) | 9,226 (36) | 25,464 |

Source: Lam and Ng (1994); *Hong Kong Energy Statistics*.
Note: Figures in brackets are percentage shares.

**Table 3.8**
**Risk Factors and Costs of Equity Capital**

| Company | Beta value | Cost of equity capital |
|---|---|---|
| CLP | 0.93 | 19.2% |
| HEC | 0.81 | 17.9% |
| HKCG | 0.96 | 19.6% |

**Table 3.9**
**Actual Rates of Return on Equity**

| Period | | CLP | HEC | HKCG |
|---|---|---|---|---|
| 1964–1978 | Nominal rate of return | 15.0% | 15.5% | 13.2% |
| | Inflation rate | 5.8% | 5.8% | 5.8% |
| | Real rate of return | 9.2% | 9.7% | 7.4% |
| 1979–1992 | Nominal rate of return | 22.4% | 20.5% | 21.9% |
| | Inflation rate | 9.4% | 9.4% | 9.4% |
| | Real rate of return | 13.0% | 11.1% | 12.5% |

**Table 3.10**
**Installed Capacity of Power Companies**

**(1) CLP**

| Year | Capacity (MW) | New (MW) | Retired (MW) | Maximum demand (MW) | Reserve capacity (%) | Remarks |
|---|---|---|---|---|---|---|
| 1952 | 67 | 20 | 3 | 45 | 50 | Units at Hok Un |
| 1953 | 67 | | | 51 | 32 | |
| 1954 | 87 | 20 | | 57 | 54 | |
| 1955 | 82 | | 5 | 71 | 16 | |
| 1956 | 82 | | | 79 | 4 | |
| 1957 | 97 | 20 | 5 | 87 | 12 | |
| 1958 | 122 | 30 | 5 | 100 | 23 | |
| 1959 | 182 | 2x30 | | 123 | 48 | |
| 1960 | 182 | | | 145 | 26 | |
| 1961 | 182 | | | 175 | 4 | |
| 1962 | 242 | 60 | | 206 | 18 | |
| 1963 | 302 | 60 | | 247 | 22 | |
| 1964 | 362 | 60 | | 297 | 22 | |
| 1965 | 362 | | | 339 | 7 | |
| 1966 | 482 | 2x60 | | 400 | 21 | |
| 1967 | 602 | 2x60 | | 455 | 32 | |
| 1968 | 662 | 60 | | 536 | 24 | |
| 1969 | 782 | 120 | | 617 | 27 | Tsing Yi Station commissioned in 1969. |
| 1970 | 870 | 120 | 12+20 | 709 | 23 | |
| 1971 | 1,110 | 2x120 | | 804 | 38 | |
| 1972 | 1,250 | 120+20 | | 869 | 44 | |
| 1973 | 1,370 | 120 | | 961 | 43 | |
| 1974 | 1,612 | 200+42 | | 1,005 | 60 | |
| 1975 | 1,812 | 200 | | 1,086 | 67 | |
| 1976 | 2,012 | 200 | | 1,232 | 63 | |
| 1977 | 2,212 | 200 | | 1,380 | 60 | |
| 1978 | 2,152 | 3x20 | | 1,572 | 37 | |
| 1979 | 2,208 | 56 | | 1,780 | 24 | |
| 1980 | 2,416 | 56+2x76 | | 1,980 | 22 | |
| 1981 | 2,650 | 4x60 | | 2,109 | 26 | Castle Peak coal-fired units |
| 1982 | 3,006 | 350 | | 2,269 | 32 | commissioned in 1982. |
| 1983 | 3,356 | 350 | | 2,475 | 36 | |
| 1984 | 3,664 | 350 | 42 | 2,652 | 38 | Hok Un Station decommissioned |
| 1985 | 3,924 | 350 | 3x30 | 2,835 | 38 | 1984–1987. |
| 1986 | 4,361 | 677 | 4x60 | 3,123 | 40 | |
| 1987 | 4,778 | 677 | 4x60+20 | 3,440 | 39 | |
| 1988 | 5,455 | 677 | | 3,593 | 52 | |
| 1989 | 5,455 | | | 3,892 | 40 | |
| 1990 | 6,132 | 677 | | 4,058 | 51 | |
| 1991 | 6,132 | | | 4,180 | 47 | |
| 1992 | 6,432 | 3x100 | | 4,365 | 47 | Penny's Bay gas turbines commissioned |
| 1993 | 6,430 | 352+150 | 504 | 4,432 | 45 | in 1992. |
| 1994 | 7,540 | 1,380+3x150 | 6x120 | 4,730 | | Capacity in China commissioned. |
| 1995 | 6,890 | 150 | 4x200 | 4,720 | 46 | Tsing Yi decommissioned in 1994–95. |

**Table 3.10 (continued)**
**Installed Capacity of Power Companies**

**(2) HEC**

| Year | Capacity (MW) | New (MW) | Retired (MW) | Maximum demand (MW) | Reserve capacity (%) | Remarks |
|---|---|---|---|---|---|---|
| 1952 | 74 | | | 45 | 65 | Units at North Point |
| 1953 | 72 | | 1.5 | 48 | 53 | |
| 1954 | 72 | | | 46 | 56 | |
| 1955 | 77 | 5 | | 57 | 35 | |
| 1956 | 92 | 15 | | 64 | 45 | |
| 1957 | 92 | | | 68 | 36 | |
| 1958 | 122 | 30 | | 75 | 63 | |
| 1959 | 152 | 30 | | 85 | 79 | |
| 1960 | 182 | 30 | | 103 | 77 | |
| 1961 | 182 | | | 114 | 60 | |
| 1962 | 182 | | | 131 | 39 | |
| 1963 | 195 | 30+10 | 3x15+12.5 | 150 | 30 | |
| 1964 | 225 | 30 | | 159 | 41 | |
| 1965 | 225 | | | 185 | 22 | |
| 1966 | 345 | 2x60 | | 201 | 72 | |
| 1967 | 345 | | | 220 | 57 | |
| 1968 | 405 | 60 | | 251 | 61 | Ap Lei Chau Station commissioned in 1968. |
| 1969 | 465 | 60 | | 274 | 70 | |
| 1970 | 465 | | | 304 | 53 | |
| 1971 | 456 | 1 | 10 | 342 | 33 | North Point Station decommissioned 1971–80. |
| 1972 | 581 | 125 | | 370 | 57 | |
| 1973 | 641 | 125 | 65 | 423 | 52 | |
| 1974 | 652 | 11 | | 442 | 48 | |
| 1975 | 777 | 125 | | 460 | 69 | |
| 1976 | 871 | 125 | 30+1 | 519 | 68 | |
| 1977 | 902 | 55 | 23+1 | 587 | 54 | |
| 1978 | 930 | 125 | 97 | 658 | 41 | |
| 1979 | 1,055 | 125 | | 691 | 53 | |
| 1980 | 1,060 | 125 | 2x60 | 795 | 33 | |
| 1981 | 1,060 | | | 852 | 24 | |
| 1982 | 1,435 | 2x250 | 125 | 932 | 54 | Lamma Station coal-fired units commissioned in 1982. |
| 1983 | 1,560 | 125 | | 1,012 | 54 | |
| 1984 | 1,685 | 125+120 | 2x60 | 1,064 | 58 | Ap Lei Chau units relocated, 1982–90. |
| 1985 | 1,685 | | | 1,150 | 47 | |
| 1986 | 1,685 | | | 1,259 | 34 | |
| 1987 | 1,915 | 350 | 120 | 1,379 | 39 | |
| 1988 | 2,005 | 350 | 2x125+10 | 1,421 | 41 | |
| 1989 | 2,005 | 4x125 | 4x125 | 1,540 | 30 | |
| 1990 | 2,255 | 2x125 | | 1,613 | 40 | |
| 1991 | 2,255 | | | 1,680 | 34 | |
| 1992 | 2,605 | 350 | | 1,819 | 43 | |
| 1993 | 2,605 | | | 1,890 | 38 | |
| 1994 | 2,605 | | | 2,021 | 29 | |
| 1995 | 2,955 | 350 | | 2,006 | 47 | |

**Table 3.10 (continued)**
**Installed Capacity of Power Companies**

**(3) HKCG**

| Year | Installed capacity (TJ) | Daily maximum demand (TJ) | Reserve capacity (%) | Remarks |
|---|---|---|---|---|
| 1970 | 6.0 | 3.6 | 67 | Units at Ma Tau Kok |
| 1971 | 6.0 | 4.0 | 50 | |
| 1972 | 6.0 | 4.4 | 36 | |
| 1973 | 9.8 | 5.2 | 88 | |
| 1974 | 12.0 | 5.4 | 122 | |
| 1975 | 12.0 | 6.9 | 74 | |
| 1976 | 12.0 | 7.3 | 64 | |
| 1977 | 12.0 | 7.8 | 54 | |
| 1978 | 15.8 | 9.1 | 74 | |
| 1979 | 15.8 | 10.4 | 52 | |
| 1980 | 15.8 | 12.9 | 22 | |
| 1981 | 27.4 | 15.6 | 76 | |
| 1982 | 38.9 | 18.4 | 111 | |
| 1983 | 38.9 | 22.4 | 74 | |
| 1984 | 38.9 | 24.4 | 59 | |
| 1985 | 61.9 | 27.8 | 23 | |
| 1986 | 61.9 | 32.2 | 92 | |
| 1987 | 112.1 | 39.0 | 187 | Tai Po Plant commissioned in 1987. |
| 1988 | 112.1 | 41.9 | 168 | |
| 1989 | 112.1 | 46.6 | 141 | |
| 1990 | 112.1 | 51.1 | 119 | |
| 1991 | 112.1 | 59.4 | 89 | |
| 1992 | 144.5 | 60.2 | 140 | |
| 1993 | 192.0 | 72.9 | 163 | |
| 1994 | 192.0 | 71.5 | 169 | |
| 1995 | 192.0 | 75.9 | 153 | |

Source: *Hong Kong Energy Statistics*, annual reports of power companies.

**Table 3.11**
**Operating Efficiency of Power Companies**

**(1) CLP**

| Year | Production (TJ) | Consumption (TJ) | System loss (TJ) | Plant factor (%) | Load factor (%) | Thermal efficiency (%) |
|---|---|---|---|---|---|---|
| 1970 | 13,335 | 11,667 | 1,675 (14) | 48 | 60 | 28.6 |
| 1971 | 14,652 | 12,889 | 1,763 (14) | 46 | 58 | 29.5 |
| 1972 | 16,190 | 14,217 | 1,973 (14) | 43 | 59 | 30.4 |
| 1973 | 17,672 | 15,745 | 1,927 (12) | 42 | 58 | 30.8 |
| 1974 | 17,356 | 15,427 | 1,929 (12) | 38 | 55 | 31.6 |
| 1975 | 18,962 | 16,682 | 2,280 (14) | 35 | 55 | 31.4 |
| 1976 | 21,690 | 19,062 | 2,628 (14) | 36 | 56 | 32.0 |
| 1977 | 24,429 | 21,462 | 2,967 (14) | 36 | 56 | 32.1 |
| 1978 | 26,870 | 23,654 | 3,216 (14) | 39 | 54 | 32.2 |
| 1979 | 30,142 | 26,790 | 3,352 (13) | 44 | 54 | 32.5 |
| 1980 | 33,353 | 29,584 | 3,769 (13) | 46 | 53 | 32.5 |
| 1981 | 34,521 | 30,886 | 3,635 (12) | 43 | 52 | 32.7 |
| 1982 | 37,672 | 33,342 | 4,330 (13) | 42 | 52 | 33.4 |
| 1983 | 41,669 | 37,730 | 3,939 (10) | 41 | 52 | 34.1 |
| 1984 | 47,007 | 42,231 | 4,776 (11) | 41 | 55 | 34.7 |
| 1985 | 51,249 | 45,474 | 5,775 (13) | 42 | 54 | 35.0 |
| 1986 | 57,311 | 50,583 | 6,728 (13) | 43 | 54 | 35.0 |
| 1987 | 63,649 | 56,409 | 7,240 (13) | 42 | 55 | 35.3 |
| 1988 | 68,470 | 60,367 | 8,103 (13) | 40 | 56 | 35.7 |
| 1989 | 73,466 | 65,014 | 8,452 (13) | 43 | 55 | 36.1 |
| 1990 | 77,520 | 68,666 | 8,854 (13) | 40 | 56 | 35.9 |
| 1991 | 86,776 | 77,182 | 9,584 (12) | 45 | 57 | 35.9 |
| 1992 | 96,175 | 86,043 | 10,132 (12) | 47 | 58 | 36.4 |
| 1993 | 98,281 | 88,108 | 10,173 (12) | 49 | 60 | 36.2 |
| 1994 | 92,205 | 81,657 | 10,548 (13) | 38 | 49 | 35.4 |
| 1995 | 93,476 | 82,649 | 10,827 (13) | 40 | 49 | 35.9 |

**Table 3.11 (continued)**
**Operating Efficiency of Power Companies**

**(2) HEC**

| Year | Production (TJ) | Consumption (TJ) | System loss (TJ) | Plant factor (%) | Load factor (%) | Thermal efficiency (%) |
|---|---|---|---|---|---|---|
| 1970 | 4,979 | 4,356 | 623 (14) | 34 | 52 | 29.1 |
| 1971 | 5,375 | 4,720 | 655 (14) | 37 | 50 | 29.3 |
| 1972 | 6,059 | 5,267 | 792 (15) | 35 | 52 | 29.8 |
| 1973 | 6,786 | 5,893 | 893 (15) | 35 | 51 | 31.7 |
| 1974 | 6,779 | 5,875 | 904 (15) | 33 | 49 | 33.5 |
| 1975 | 7,499 | 6,444 | 1,055 (16) | 31 | 52 | 35.3 |
| 1976 | 8,251 | 7,128 | 1,123 (16) | 30 | 50 | 35.7 |
| 1977 | 9,493 | 8,330 | 1,163 (14) | 33 | 51 | 35.4 |
| 1978 | 10,404 | 9,140 | 1,264 (14) | 36 | 50 | 35.9 |
| 1979 | 10,825 | 9,680 | 1,145 (12) | 33 | 50 | 35.9 |
| 1980 | 12,143 | 10,843 | 1,300 (12) | 37 | 48 | 35.7 |
| 1981 | 13,344 | 11,585 | 1,759 (15) | 40 | 44 | 35.7 |
| 1982 | 14,501 | 12,406 | 2,095 (17) | 39 | 44 | 35.3 |
| 1983 | 17,629 | 14,119 | 3,510 (25) | 37 | 42 | 34.9 |
| 1984 | 17,517 | 14,598 | 3,451 (24) | 33 | 45 | 35.1 |
| 1985 | 18,049 | 15,646 | 2,351 (15) | 34 | 48 | 34.8 |
| 1986 | 19,797 | 17,359 | 2,413 (14) | 37 | 50 | 34.8 |
| 1987 | 21,862 | 19,120 | 2,745 (14) | 39 | 50 | 35.3 |
| 1988 | 23,358 | 20,452 | 2,806 (14) | 37 | 52 | 36.1 |
| 1989 | 25,042 | 21,946 | 3,096 (14) | 40 | 51 | 36.2 |
| 1990 | 26,736 | 23,605 | 6,131 (13) | 38 | 53 | 36.0 |
| 1991 | 28,024 | 24,977 | 3,047 (12) | 39 | 50 | 35.7 |
| 1992 | 30,097 | 25,974 | 4,123 (16) | 38 | 48 | 36.0 |
| 1993 | 32,736 | 27,904 | 4,832 (17) | 40 | 54 | 36.0 |
| 1994 | 33,772 | 29,725 | 4,047 (14) | 41 | 53 | 35.7 |
| 1995 | 34,185 | 30,168 | 4,017 (13) | 42 | 54 | 35.3 |

**Table 3.11 (continued)**
**Operating Efficiency of Power Companies**

**(3) HKCG**

| Year | Production Consumption (TJ) | Plant factor (%) | Load factor (%) |
|---|---|---|---|
| 1970 | 958 | 44 | 73 |
| 1971 | 1,035 | 47 | 71 |
| 1972 | 1,135 | 52 | 70 |
| 1973 | 1,278 | 36 | 67 |
| 1974 | 1,476 | 34 | 75 |
| 1975 | 1,609 | 37 | 64 |
| 1976 | 1,891 | 43 | 71 |
| 1977 | 2,077 | 47 | 73 |
| 1978 | 2,503 | 43 | 75 |
| 1979 | 3,021 | 52 | 80 |
| 1980 | 3,524 | 61 | 75 |
| 1981 | 4,034 | 40 | 71 |
| 1982 | 4,858 | 34 | 72 |
| 1983 | 5,924 | 42 | 72 |
| 1984 | 6,907 | 49 | 77 |
| 1985 | 7,979 | 35 | 79 |
| 1986 | 9,043 | 40 | 77 |
| 1987 | 10,584 | 26 | 74 |
| 1988 | 12,247 | 30 | 80 |
| 1989 | 13,671 | 33 | 80 |
| 1990 | 15,056 | 37 | 81 |
| 1991 | 16,238 | 40 | 75 |
| 1992 | 18,207 | 35 | 83 |
| 1993 | 19,198 | 27 | 72 |
| 1994 | 20,727 | 30 | 79 |
| 1995 | 21,972 | 31 | 79 |

Source: *Hong Kong Energy Statistics.*
Note:
1. Plant factor is average daily production divided by installed capacity.
2. Load factor is average daily production divided by peak demand within the year.
3. Figures in brackets are system loss as a percentage of consumption.
4. As from 1993, production of CLP includes electricity from China.

**Table 3.12**

**Capital Expenditures (in HK$ million) of Power Companies, 1964–1995**

| | CLP | | | HEC | HKCG |
|---|---|---|---|---|---|
| Year | Generation | Trans-mission | Total | | |
| 1964 | N.A. | N.A. | 92 | 53 | |
| 1965 | 45 | 80 | 125 | 65 | |
| 1966 | 95 | 96 | 191 | 67 | |
| 1967 | 87 | 64 | 151 | 90 | |
| 1968 | 72 | 61 | 133 | 64 | |
| 1969 | 74 | 81 | 155 | 37 | |
| 1970 | 81 | 84 | 165 | 53 | |
| 1971 | 88 | 135 | 223 | 120 | |
| 1972 | 123 | 149 | 272 | 90 | |
| 1973 | 125 | 112 | 237 | 112 | |
| 1974 | 121 | 131 | 252 | 133 | |
| 1975 | 112 | 137 | 249 | 150 | |
| 1976 | 88 | 127 | 215 | 160 | |
| 1977 | 42 | 162 | 204 | 276 | |
| 1978 | 176 | 172 | 348 | 250 | |
| 1979 | 780 | 373 | 1,153 | 536 | |
| 1980 | 1,623 | 812 | 2,435 | 943 | |
| 1981 | 1,671 | 1,186 | 2,857 | 1,639 | |
| 1982 | 2,045 | 913 | 2,958 | 1,439 | |
| 1983 | 2,593 | 818 | 3,411 | 712 | |
| 1984 | 2,842 | 889 | 3,731 | 765 | |
| 1985 | 2,548 | 1,109 | 3,657 | 1,402 | |
| 1986 | 2,321 | 1,149 | 3,470 | 1,764 | 610 |
| 1987 | 1,572 | 1,149 | 2,721 | 1,106 | 296 |
| 1988 | 813 | 1,351 | 2,164 | 1,798 | 328 |
| 1989 | 868 | 1,892 | 2,760 | 2,433 | 456 |
| 1990 | 581 | 3,344 | 3,925 | 2,439 | 1,050 |
| 1991 | 2,584 | 3,516 | 6,100 | 2,667 | 954 |
| 1992 | 987 | 2,982 | 3,969 | 2,970 | 816 |
| 1993 | 3,692 | 2,785 | 6,477 | 3,486 | 693 |
| 1994 | 4,465 | 2,943 | 7,408 | 4,668 | 774 |
| 1995 | 5,319 | 3,396 | 8,715 | 5,459 | 786 |
| Total: | **37,304** | 30,607 | **67,911** | **36,226** | **6,763** |
| | ------ (1979–1995) ------ | | | (1979–1995 | (1986-1995) |

Source: Annual reports of CLP, HEC and HKCG.

# Bibliography


1. ANR Pipeline Company (1994). *Natural gas: the Fuel, the Future, the Regulation.*

2. Armstrong, Mark, S. Cowan, and J. Vickers (1994). *Regulatory Reform: Economic Analysis and British Experience.* Cambridge: The MIT Press.

3. Asch, P., and R. S. Seneca (1984). *Government and the Market Place.* 2nd edition, Orlando: The Dryden Press.

4. Averch, H., and L. L. Johnson (1962). "Behaviour of the Firm under Regulatory Constraint." *American Economic Review*, 52:1052–1069.

5. Bailey, E. E., and R. D. Coleman (1971). "The Effect of Lagged Regulation in an Averch-Johnson Model." *Bell Journal of Economics and Management Science*, 2:278–292.

6. Baumol, W. J., and A. K. Klevorick (1970). "Input Choices and Rate-of-return Regulation: An Overview of the Discussion." *Bell Journal of Economics and Management Science*, 2:162–190.

7. Beesley, M. E., and S. C. Littlechild (1989). "The Regulation of Privatized Monopolies in the United Kingdom." *Rand Journal of Economics*, 20: 454–472.

8. Berry, David (1989). "US Cogeneration Policy in Transition." *Energy Policy*, October: 471–484.

9. British Gas Corporation International Consultancy Service (1981). *Report on the Safety and Legal Aspects of Both Town Gas and LPG Operations in Hong Kong.*

10. Cameron, Nigel (1982). *Power: The Story of China Light.* Hong Kong: Hong Kong Oxford University Press.

11. China Light & Power Company Limited (CLP). *Annual Report.* 1946–1995.

12. China Light & Power Company Limited (CLP) (1964). "Proposed Scheme of Control". in *Annual Report.*

13. Chou, Win-lin (1979). "Energy Consumption in Hong Kong." *Economic Journal*, 1979:32–49.

14. Chow, Larry C. H. (1989). "Utilities." In *The Other Hong Kong Report 1989*, edited by Tsim, T. L., and Bernard H. K. Luk, Hong Kong: The Chinese University Press.

15. Coates, Austin (1977). *A Mountain of Light: The Story of the Hongkong Electric Company*. London: Heineman.

16. Crew, Michael A., and P. R. Kleindorfer (1987). "Productivity Incentives and Rate-of-return Regulation." In *Regulating Utilities in an Era of Deregulation*, edited by Michael A. Crew, pp. 7–23, London: Macmillan Press.

17. Crew, A. M., P. R. Kleindorfer, and D. L. Schlenger (1987). "Governance Costs of Regulation for Water Supply." In *Regulating Utilities in an Era of Deregulation*, edited by Michael A. Crew, pp. 43–62, London: Macmillan Press.

18. Crocker, Keith J, and Scott E. Masten (1996). "Regulation and Administrated Contracts Revisited: Lessons from Transaction-Cost Economics for Public Utility Regulation." *Journal of Regulatory Economics*, 9:5–39.

19. Economic Services Branch (Hong Kong government) (1996). *Government Response to Consumer Council's Report on Assessing Competition in the Domestic Water Heating and Cooking Fuel Market*. February 1996.

20. Ernst & Whinney (1984). *Consultancy to Review the Government's Monitoring Arrangement of the Power Companies — Prepared for Government of Hong Kong.*

21. Fox-Penner, Peter S. (1990). "Cogeneration After PURPA: Energy Conservation and Industry Structure." *Journal of Law and Economics*, 33:517–552.

22. Friends of the Earth (1996). *Power Without Policy: Friends of the Earth's Position on Energy Policy*. Hong Kong: Friends of the Earth, October 1996.

23. Gilbert, R. J., and D. M. Newbery (1994). "The Dynamic Efficiency of Regulatory Constitutions." *Rand Journal of Economics*, 25:538–554.

24. Gorak, T. C., and Ray, D. J. (1995). "Efficiency and Equity in the Transition to a New Natural Gas Market." *Land Economics*, 71:368–385.

25. Hall, Christopher D (1996). *The Uncertain Hand: Hong Kong Taxis and*

*Tenders*. Hong Kong: The Chinese University Press.

26. Hempling Scott (1995). "Electric Utility Holding Companies: The New Regulatory Challenges." *Land Economics*, 71:343–353.

27. Hongkong Electric Company Limited (HEC). *Annual Report and Accounts*. 1955–1975.

28. Hongkong Electric Holdings Limited (HEC). *Annual Report*. 1976–1995.

29. Hong Kong Census and Statistics Department (1990). "Energy Statistics in Hong Kong." In *Hong Kong Monthly Digest of Statistics*, pp.101–108, January 1990.

30. Hong Kong Census and Statistics Department (1994). "Hong Kong Energy Statistics, 1983–93." in *Hong Kong Monthly Digest of Statistics*, pp.110–117, December 1994.

31. Hong Kong China & Gas Company Limited (HKCG). *Annual Report*. 1963–1995.

32. Hong Kong Consumer Council (1995). *Assessing Competition in the Domestic Water Heating and Cooking Fuel Market*.

33. Hong Kong government. *Hong Kong Annual Report*, chapter on public utilities, 1946–1996.

34. Hong Kong government (1959). *Electricity Supply Companies Commission — Report*.

35. Hong Kong government (1982). *The Schemes of Control*.

36. Hong Kong government, and CLP (1992). *The Scheme of Control Agreement*.

37. Hutchinson, Mark, Ming Fang, and Gary Heinke (1995). "The Environmental Impact of Domestic Hot Water Consumption in Hong Kong." *Asia Engineer*, November 1995, pp. 19–28.

38. International Energy Agency (IEA) (1992). *Electricity Supply in the OECD*. Paris: IEA.

39. International Energy Agency (IEA) (1994). *Energy in Developing Countries*. Paris: IEA.

40. Joskow, Paul L. (1974). "Inflation and Environmental Concern: Structural Change in the Process of Public Utility Price Regulation." *Journal of Law and Economics*, 17:291–327.

41. Joskow, Paul L. (1986). "Incentive Regulation for Electric Utilities." *Yale Journal on Regulation*, 4:1-49.

42. Joskow, Paul L., and Richard Schmalensee (1983). *Markets for Power: An Analysis of Electric Utility Deregulation*. Cambridge: The MIT Press.

43. Kwong, Kai-sun (1995). "Infrastructure." In *The Other Hong Kong Report 1985*, edited by Stephen Y. L. Cheung, and Stephen M. H. Sze. Hong Kong: The Chinese University Press.

44. Lam, J. C., and Ng, A. K. W. (1994). "Energy Consumption in Hong Kong." *Energy*, 19:1157–1164.

45. Lam, Pun-lee (1996a). "Restructuring the Hong Kong Gas Industry." *Energy Policy*, 24:713–722

46. Lam, Pun-lee (1996b). "Transition to Competition in Hong Kong's Local Telephone Industry," *Telecommunications Policy*, 20:517–529.

47. Lam, Pun-lee (1996c). *The Scheme of Control on Electricity Companies. Hong Kong: The Chinese University Press.*

48. Liu, Pak-wai (1990). "Utilities and Telecommunications." In *The Other Hong Kong Report 1990*, edited by Richard Y. C. Wong, and Joseph Y. S. Cheng. Hong Kong: The Chinese University Press.

49. Liu, Pak-wai (1991). "Utilities and Telecommunications: Regulation of Monopolies." In *The Other Hong Kong Report 1991*, edited by Yun-wing Sung, and Ming-kwan Lee. Hong Kong: The Chinese University Press.

50. Lyon, Thomas P. (1995). "Regulatory Hindsight Review and Innovation by Electric Utilities." *Journal of Regulatory Economics*, 7:233–254.

51. Norton, Seth W (1985). "Regulation and Systematic Risk: The Case of Electric Utilities." *Journal of Law and Economics*,28:671–686.

52. Priest, George L. (1993). "The Origins of Utility Regulation and the 'Theories of Regulation' Debate." *Journal of Law and Economics*, 36: 289–323.

53. Reddy, Sudhakara (1996). "Economic Evaluation of Demand-Side Management Options Using Utility Avoided Costs." *Energy*, 21:473–482.

54. Robison, H. David, Wallace N. Davidson III, and John L. Glascock (1995). "The Formation of Public Utility Holding Companies and Their Subsequent Diversification Activity." *Journal of Regulatory Economics*, 7:199–214.

55. Rusco, F. W., and W. D. Walls (1995). *Clearing the Air: Vehicular Emissions Policy for Hong Kong*. Hong Kong: The Chinese University Press.

56. Shleifer, Andrei (1985). "A Theory of Yardstick Competition." *Rand Journal of Economics*, 16:219–327.

57. Stalon, Charles G. (1995). "Restructuring the U.S. Electric Industry for the 21st Century." Lecture notes for the Annual Regulatory Studies Program at Michigan State University.

58. Summerton, Janes, and Ted K. Bradshaw (1991). "Towards a Dispersed Electrical System: Challenges to the Grid." *Energy Policy*, January/February 1991, pp. 24–34.

59. Vickers John, and George Yarrow (1988). "The Energy Industries." *Privatization: An Economic Analysis*, Cambridge: The MIT Press, pp. 243–340

60. Vickers, John, and George Yarrow (1991a). "The British Electricity Experiment." *Economic Policy: An European Forum*, 12:188–231.

61. Vickers, John, and George Yarrow (1991b). "Economic Perspectives on Privatization." *Journal of Economic Perspectives*, 5:111–132.

62. Yarrow, George (1994). "Privatization, Restructuring, and Regulatory Reform in Electricity Supply." In *Privatization & Economic Performance*, edited by Matthew Bishop, John Kay, and Colin Mayer, Oxford: Oxford University Press, pp. 62–88,

# Index

## O



## P



## Q



## R



## S



## T

# About the Author

Pun-Lee Lam received his PhD degree in economics from the University of Bristol in 1995. He is currently Assistant Professor in the Department of Business Studies at the Hong Kong Polytechnic University. His research interest lies in the area of government regulation of public utilities. Dr. Lam's dissertation is on the government control of electricity companies and a portion of it has been published as a monograph entitled *The Scheme of Control on Electricity Companies*. Results of his research on government policies relating to utilities have been published in various international journals that address policy issues.

# The Hong Kong Economic Policy Studies Series

| *Titles* | *Authors* |
|---|---|
| **Hong Kong and the Region** | |
| ❑ The Political Economy of Laissez Faire | Yue-Chim Richard WONG |
| ❑ Hong Kong and South China: The Economic Synergy | Yun-Wing SUNG |
| ❑ Inflation in Hong Kong | Alan K. F. SIU |
| ❑ Trade and Investment: Mainland China, Hong Kong and Taiwan | K. C. FUNG |
| ❑ Tourism and the Hong Kong Economy | Kai-Sun KWONG |
| ❑ Economic Relations between Hong Kong and the Pacific Community | Yiu-Kwan FAN |
| **Money and Finance** | |
| ❑ The Monetary System of Hong Kong | Y. F. LUK |
| ❑ The Linked Exchange Rate System | Yum-Keung KWAN<br>Francis T. LUI |
| ❑ Hong Kong as an International Financial Centre: Evolution, Prospects and Policies | Y. C. JAO |
| ❑ Corporate Governance: Listed Companies in Hong Kong | Richard Yan-Ki HO<br>Stephen Y. L. CHEUNG |
| ❑ Institutional Development of the Insurance Industry | Ben T. YU |
| **Infrastructure and Industry** | |
| ❑ Technology and Industry | Kai-Sun KWONG |
| ❑ Competition Policy and the Regulation of Business | Leonard K. CHENG<br>Changqi WU |
| ❑ Telecom Policy and the Rise of Digital Media | Milton L. MUELLER |
| ❑ Port Facilities and Container Handling Services | Leonard K. CHENG<br>Yue-Chim Richard WONG |

| *Titles* | *Authors* |
| --- | --- |
| ❑ Efficient Transport Policy | Timothy D. HAU<br>Stephen CHING |
| ❑ Competition in Energy | Pun-Lee LAM |
| ❑ Privatizing Water and Sewage Services | Pun-Lee LAM<br>Yue-Cheong CHAN |

## Immigration and Human Resources

| | |
| --- | --- |
| ❑ Labour Market in a Dynamic Economy | Wing SUEN<br>William CHAN |
| ❑ Immigration and the Economy of Hong Kong | Kit Chun LAM<br>Pak Wai LIU |
| ❑ Youth, Society and the Economy | Rosanna WONG<br>Paul CHAN |

## Housing and Land

| | |
| --- | --- |
| ❑ The Private Residential Market | Alan K. F. SIU |
| ❑ On Privatizing Public Housing | Yue-Chim Richard WONG |
| ❑ Housing Policy for the 21st Century: Homes for All | Rosanna WONG |
| ❑ Financial and Property Markets: Interactions Between the Mainland and Hong Kong | Pui-King LAU |
| ❑ Town Planning in Hong Kong: A Critical Review | Lawrence Wai-Chung LAI |

## Social Issues

| | |
| --- | --- |
| ❑ Retirement Protection: A Plan for Hong Kong | Francis T. LUI |
| ❑ Income Inequality and Economic Development | Hon-Kwong LUI |
| ❑ Health Care Reform: An Economic Perspective | Lok-Sang HO |
| ❑ Economics of Crime and Punishment | Siu Fai LEUNG |